# THE ULTIMATE NAVY SEALS BUG-IN GUIDE

## HOW TO MAKE YOUR HOUSE THE SAFEST PLACE USING PROVEN MILITARY STRATEGIES IN ANY CRISIS

### JAKE RYDER

# CONTENTS

| | |
|---|---|
| Introduction | vii |
| **1. UNDERSTANDING YOUR HOME'S VULNERABILITIES** | 1 |
| Conducting a Home Safety Audit | 1 |
| Evaluating Landscape Features for Potential Risks | 3 |
| Utilizing Shelter-in-Place Strategies to Protect Your Family and Home | 4 |
| Identifying Weak Points in Doors and Windows | 6 |
| Assessing Electrical and Utility Vulnerabilities | 7 |
| Summary and Reflections on Understanding Your Home's Vulnerabilities | 9 |
| **2. SECURING THE INTERIOR AND EXTERIOR OF YOUR HOME** | 10 |
| Installing Durable Perimeter Fencing | 10 |
| Upgrading Door and Window Locks | 12 |
| Strategic Outdoor Lighting Placement | 13 |
| Setting up Security Cameras and Alarms | 14 |
| Reinforcing Internal Doors With Deadbolts | 16 |
| Concluding Thoughts | 17 |
| **3. DEVELOPING EFFECTIVE EMERGENCY PLANS** | 18 |
| Establishing Communication Protocols | 18 |
| Planning Escape Routes and Meeting Points | 20 |
| Designating Roles and Responsibilities | 21 |
| Practicing Drills and Scenario Responses | 22 |
| Incorporating Technology in Emergency Planning | 24 |
| Concluding Thoughts | 26 |
| **4. ASSEMBLING ESSENTIAL RESOURCES—FOOD, WATER, AND OFF-GRID ENERGY SOLUTIONS** | 27 |
| Compiling Food and Water Reserves | 28 |
| Food Stockpiling & Storage | 29 |
| Water Sources & Purification | 32 |
| Off-Grid Energy Solutions | 36 |
| Communication Devices and Power Sources | 38 |
| Concluding Thoughts | 39 |
| **5. MEDICAL PREPAREDNESS—ADDRESSING MEDICAL EMERGENCIES AND CARE** | 40 |
| First Aid and Medical Supply Considerations | 41 |
| Training in First Aid Techniques | 42 |
| Management of Over-The-Counter and Prescription Medications | 43 |
| Natural Remedies and Self-Care for Common Ailments | 47 |

    Special Considerations for Families and Unique Needs      48
    Summary and Reflections      50

6. **ESSENTIAL SURVIVAL GEAR AND PROTECTIVE EQUIPMENT**      51
    Basic Survival Gear Essentials      51
    Personal Protective Equipment (PPE)      53
    Customized Equipment Based on Risk      54
    Maintaining and Checking Gear Readiness      55
    Essential Survival Gear Recommended by Elite Military Experts      56
    Final Thoughts      57

7. **FINANCIAL AND LEGAL PREPAREDNESS FOR CRISIS SITUATIONS**      59
    Managing Finances During a Crisis      59
    Protecting Your Assets      61
    Cash, Digital Currency, and Bartering      62
    Insights and Implications      63

8. **PSYCHOLOGICAL PREPAREDNESS AND RESILIENCE**      65
    Understanding Your Stress Responses During Emergencies      65
    Building and Encouraging Group Trust      66
    Techniques for Staying Calm Under Pressure      68
    Mindfulness and Mental Training Exercises      69
    Decision-Making Strategies in High-Stress Scenarios      71
    Final Insights      72

9. **LEVERAGING MILITARY TACTICS FOR HOME DEFENSE**      73
    Camouflage and Concealment Techniques      73
    Set up Control Points      74
    Use *Fieldcraft* Skills      76
    Implement Patrols and Surveillance      77
    Utilizing Decoys and Distractions      78
    Final Thoughts      79

10. **BUILDING A SUPPORTIVE COMMUNITY NETWORK**      81
    Benefits of Community Watch Programs      82
    Creating Community Watch Programs      83
    Establishing Mutual Aid Agreements      85
    Coordinating Neighborhood Emergency Plans      86
    Sharing Resources and Skills      87
    Hosting Regular Safety Meetings      88
    Reflection      90

11. **ADAPTING TO DIFFERENT CRISIS SCENARIOS**      91
    Preparing for Natural Disasters      91
    Considering Pandemics and Health Crises      92
    Dealing With Long-Term Power Outages      94
    How to Deal With Hazardous Materials      95
    Bringing It All Together      96

12. **LONG-TERM MAINTENANCE AND IMPROVEMENT** 97
    Scheduled Safety Audits and Updates 97
    Review and Replenish Supplies 99
    Staying Informed on Emerging Threats 100
    Incorporating New Technologies and Innovations 101
    How to Create Checklists 102
    Final Thoughts 105

    Conclusion 107
    References 111

© **Copyright 2024 - All rights reserved.**

The content contained within this book may not be reproduced, duplicated or transmitted without direct written permission from the author or the publisher.

Under no circumstances will any blame or legal responsibility be held against the publisher, or author, for any damages, reparation, or monetary loss due to the information contained within this book, either directly or indirectly.

Legal Notice:

This book is copyright protected. It is only for personal use. You cannot amend, distribute, sell, use, quote or paraphrase any part, or the content within this book, without the consent of the author or publisher.

Disclaimer Notice:

Please note the information contained within this document is for educational and entertainment purposes only. All effort has been executed to present accurate, up to date, reliable, complete information. No warranties of any kind are declared or implied. Readers acknowledge that the author is not engaged in the rendering of legal, financial, medical or professional advice. The content within this book has been derived from various sources. Please consult a licensed professional before attempting any techniques outlined in this book.

By reading this document, the reader agrees that under no circumstances is the author responsible for any losses, direct or indirect, that are incurred as a result of the use of the information contained within this document, including, but not limited to, errors, omissions, or inaccuracies.

# INTRODUCTION

Today, the safety of your home has become a priority like never before. Your home should be a sanctuary, a place where you walk through the door and leave your worries behind, easing you into an environment that makes you feel comfortable and secure. For many homeowners and renters, this sense of security is not just about locking doors, as it touches every aspect of daily life and family well-being. In a world filled with uncertainties, the assurance of a secure home provides you with peace of mind.

When you return home after a long day of stress from work and daily responsibilities, the moment you step inside, you are greeted by a secure space that allows your worries to be put to rest. Living in constant fear can be draining, so having that assurance from the safety measures explained in this book is freeing.

The threats to home safety often include risks most of us may overlook. These unnoticed hazards, though seemingly minor, could escalate into catastrophic problems if left unaddressed. Recognizing and identifying these potential dangers is the first crucial step toward securing our homes. It empowers us to take precautionary actions, transforming potential threats into opportunities for safeguarding.

Furthermore, preparedness is truly essential for everyone. When you're prepared, you're able to calmly tackle any crisis that may come your way. For example, knowing exactly where emergency supplies are located turns panic into practicality. However, preparedness doesn't merely prepare us for the worst; it allows for steady management of any unexpected event, turning what could be chaotic spirals into organized, manageable challenges.

While traditional methods of ensuring safety suffice for everyday scenarios, using strategies from military tactics offers a unique, disciplined approach to home defense. Approaching a

## INTRODUCTION

crisis with strategic military precision ensures optimal security in the face of danger. Having strategic planning skills elevates the safety of your home from being merely adequate to formidably protected. Subsequently, the empowerment derived from knowing your surroundings are defended to your best capabilities instills unmatched confidence and offers peace of mind that extends beyond physical barriers.

On the other hand, although you may be in charge of keeping your house safe, no fortress stands alone. Community involvement plays a big role in enhancing the safety of you and your family. A neighborhood bound together by mutual vigilance, where families rally to support and watch over each other, means that they can not only help you, but you can also help them with your thorough strategies. Each community member, including you, can actively participate in a collective force against crime and danger. Together, you are safer and more likely to experience noticeable results on a wider scale.

This book is your guide to understanding and implementing comprehensive safety measures that cater to you and your family's practical needs and emotional reassurance. It uncovers more than manual locks and alarms, as it discusses the psyche of security, exploring the profound impact of feeling safe at home. Across these chapters, you will discover safety strategies and techniques drawn from various disciplines, offering practical ways to approach different scenarios and threats. By starting this book, you join a journey toward transforming your home into a bastion of safety for yourself and your loved ones.

Whether you are a homeowner vigilant about family well-being, an enthusiast keen on blending survival skills with everyday living, or a parent determined to protect and equip your family for possible crises, this book offers invaluable knowledge and tools. It showcases the intricacies of home safety with clarity and authority, ensuring that you know how to tackle the vulnerabilities within your personal space head-on.

With systematic approaches and practical advice, together, we will unpack the elements of a secure home, redefining what it means to live in safety. Through a blend of military strategies and straightforward instructions, you'll learn how to elevate your home's defenses, turning it into a resilient stronghold.

Through lots of practice and learning, you'll be able to turn preparedness into a lived reality. You will find yourself enriched with a better mindset and practice to handle whatever uncertainties life throws your way. Here begins a new chapter of your life in safeguarding what matters most: the people and places you hold dear. Welcome to a path toward not only protecting but truly celebrating the sanctity of your home.

# 1

# UNDERSTANDING YOUR HOME'S VULNERABILITIES

Most of us have weak points in our homes that we're completely unaware of. These vulnerabilities can pose serious threats in the event of a crisis or emergency. Identifying and understanding the vulnerabilities within your home's structure and surroundings helps you create a secure environment.

Every home has its unique set of strengths and weaknesses that can be influenced by various factors such as location, design, and existing security measures. Making proactive and thorough assessments allows you to prepare yourself to handle emergencies effectively. In this chapter, we will discuss how to recognize these weak points, emphasizing the importance of vigilance and regular inspection. Taking preventative action can provide you with great peace of mind.

By learning how to assess the safety of your home with safety audits, you will be able to minimize the vulnerabilities in your home that could be targeted by intruders. The structural aspects of your home, landscaping, and exterior features can all have an impact on the overall security of your home. The chapter also covers the integration of modern technology to improve your safety measures, as well as the value of involving family members in the process to educate and prepare every member of the household. These topics will provide you with foundational knowledge for addressing your home's defenses against various threats.

## Conducting a Home Safety Audit

Before taking tactical approaches to enhancing the safety of your home, you should start by conducting a home safety audit. Doing this helps you identify the vulnerabilities you should prepare. In addition, this process allows you to generate a road map for enhancing your protection against potential threats.

The following steps can help you conduct a home safety audit that informs you and your family of the best way forward:

1. **Comprehensive evaluation:** Begin with a comprehensive evaluation of all entry points to your home. Doors and windows are common access routes for intruders, so it's crucial to check them thoroughly for any weaknesses or unsecured areas. Regularly inspect door frames to ensure they're sturdy and free from damage, as even minor compromises can be exploited by burglars. Secure doors with quality locks, and consider adding deadbolts for extra protection. Windows should also be assessed; modern locking systems and impact-resistant glass can deter unlawful entry. It's also important to consider whether there is visibility around these entrances through adequate lighting and trimming shrubs or trees that could provide cover for intruders.
2. **Interior focus:** Next, turn your attention to the interior of your home, focusing on areas like garages and basements. These spaces often lack reliable security measures, making them attractive targets for burglars. Verify that your garage doors are equipped with secure locks and cannot be easily forced open. If your garage connects to your home, ensure the interior door is as secure as the front entrance. Moreover, your basement may have windows or exterior doors vulnerable to tampering. Install window bars or grates and reinforce basement doors to bolster security.
3. **Thorough checklist:** A structured checklist is an invaluable tool you can use to stay on track with your safety audit. It serves as a guide to ensure no area is neglected and every potential risk is addressed. Start by listing all entry points and the specific features that require inspection, such as lock mechanisms, door material, and window seals. Include a section dedicated to interior spaces, noting particular security enhancements needed. Regularly updating this checklist to reflect improvements or newly discovered vulnerabilities is recommended. When you systematically work through this list, you create a comprehensive picture of your home's security status and develop a plan for addressing any issues.
4. **Family involvement:** Involving family members in the safety audit process fosters a collaborative approach to home security. Assigning roles and responsibilities can make the task less daunting and more efficient. Children, for instance, can be tasked with checking that windows are closed and locked at night or helping test alarms. Encouraging this can help your children build healthy habits that they reinforce as they grow older, prioritizing safety in your home and their future homes as well. Adults can take charge of inspecting doors and reviewing the checklist. Engaging everyone in the household makes this less overwhelming while

allowing less room for error, as multiple eyes can identify any weak points that go unnoticed.

5. **Modern technology:** Utilizing modern technology can further streamline the home audit process. Smart home devices, such as video doorbells and surveillance cameras, provide real-time monitoring and alerts, augmenting traditional methods of securing entry points. Motion sensors and smart locks add layers of security that can be managed remotely via smartphones, offering peace of mind whether you're home or away.
6. **External resources:** Don't overlook the benefits of external resources. Many local fire departments, for instance, offer free home safety checks to assess various hazards and make recommendations for improvement (*Complete Home Safety*, 2022). Local paid security services can also enhance your peace of mind. Taking advantage of these services can supplement your own efforts, providing an expert perspective that strengthens your overall assessment.

Though these steps might seem extensive, remember that ensuring the safety of your home is an ongoing commitment. Regular audits help accommodate changes in living situations, address wear and tear on security fixtures, and adapt to evolving security threats.

Evaluating Landscape Features for Potential Risks

The security of your home doesn't begin and end with locked doors and windows. Your outdoor spaces are just as important as the interior aspects of your home. A well-kept landscape can promote a sense of security, as well as deter attention from potential threats. The following tips can help you optimize your exterior landscape's safety:

- **Minimize big landscape features:** Often, the beauty of a lush garden or a towering fence might inadvertently provide cover for potential intruders. These landscape features can obscure sight lines, making it harder to detect any unusual activity around your property. Shrubs and trees planted too close to windows or entrances can serve as optimal hiding spots for unwanted visitors. Therefore, consider trimming down overgrown plants and repositioning large decorative elements to improve visibility. Keep fences at a level where you can still see over them from a distance, or install lattice tops that maintain privacy while allowing light and sight through. This strategic approach ensures that no one can easily move around without being noticed.
- **Clear your walkways:** In addition to potentially obscured sight lines, pathways and walkways around your home often present another set of challenges: tripping hazards. Unkempt paths cluttered with debris, uneven stones, or worn-out surfaces can make evacuation during emergencies dangerous and slow. Identifying and

rectifying these hazards is essential for maintaining a safe environment. Regularly inspect all outdoor routes for loose gravel, protruding tree roots, or broken pavement. Simple actions like repairing cracks or installing anti-slip surfaces in high-traffic areas can greatly enhance safety. Adding path lights illuminates these hazards and also serves as a deterrent to intruders by highlighting entry points around your home (Edwards, R., 2022).

- **Maintain your landscape:** Regularly maintaining your garden plays a significant role in your home's security. Neglected gardens or backyards may attract intruders. Thus, regular gardening not only enhances curb appeal but also demonstrates vigilance. When your lawn is well-maintained, it suggests an active presence, which can dissuade potential intruders who prefer unkempt environments where they won't be easily spotted. Routine pruning, mowing, and cleaning ensure your outdoor space remains open and transparent, reinforcing safety through diligent upkeep.
- **Add natural barriers:** Incorporating natural barriers into your outdoor design is a clever way to bolster security. Natural features such as large rocks, mature trees, and varying elevations can be leveraged as part of your defensive plan. Position trees or large shrubberies near perimeter fences to create a buffer zone. This deters intrusion by creating natural obstacles and promotes the strategic use of features in your landscape.
- **Utilize outdoor features to your advantage:** Beyond creating barriers, it's important to utilize these features strategically. For instance, planting thorny bushes under first-floor windows offers an additional layer of defense, making it less attractive for anyone attempting unauthorized access. Trees can also house lighting or security cameras, extending surveillance capabilities beyond immediate perimeter walls. According to Pogue (2023), integrating solar-powered lights with these natural structures provides an eco-friendly and effective method to illuminate dark spots, further enhancing visibility while conserving energy.

Remain vigilant over your landscape. As landscapes change with growth and seasons, adjustments may be necessary to maintain optimal security levels. Pruning your landscape will encourage you to bring this attention to detail inside your home.

Utilizing Shelter-in-Place Strategies to Protect Your Family and Home

In times of crisis, the decision to stay inside rather than evacuate can significantly impact your safety and well-being. Staying put, or *staying in*, is often a strategic choice that combines security, resource management, and adaptability to unfolding situations. Here are some advantages that showcase why staying in is such a valuable decision:

- **Traveling during a crisis can be dangerous:** Traveling during a crisis introduces numerous hazards. Roads may become congested, leading to long delays and increased stress. Emergency situations can also create unpredictable conditions with rapidly changing circumstances such as weather or potential violence. By choosing to stay inside, you avoid these dangers. In addition, vehicles are vulnerable to external threats and breakdowns while sheltering in your home reduces the risks of accidents or being caught in hazardous environments.
- **Staying home provides you with more control:** When you decide to stay home, you have better control over who enters your space. This is crucial during emergencies when public order might be compromised. Securing entry points like doors and windows increases your home's security, minimizing the risk of intruders. Reinforced locks and barricades can serve as the first line of defense, allowing you to maintain a secure perimeter for yourself and your loved ones. Staying in also allows for constant monitoring of news updates and communication with authorities using radios or smartphones, ensuring you are informed about any changes or directives. Staying informed helps you make sound decisions based on credible information rather than rumors or assumptions circulating outside.
- **Staying in promotes effective resource management:** Another advantage of staying in is effective resource management. When you stay in one location, all necessary supplies are concentrated, eliminating the logistical challenges of transporting essentials. In this context, a well-prepared emergency kit should include food, water, first aid supplies, and personal hygiene items to last several days. Having everything in one place ensures that you're not scrambling for resources amid chaos. This strategy is particularly beneficial for families or those with dependents, making it easier to meet everyone's needs.
- **Preparing in advance reinforces psychological readiness.** In the event of an emergency or crisis, having psychological readiness is imperative. Maintaining a calm mindset helps you manage the stresses associated with crises. Moreover, preparing mentally for different scenarios reduces panic and enhances response capabilities. Preparation also involves testing and rehearsing different emergency plans. Therefore, conduct regular drills to familiarize everyone with the processes, ensuring no steps are overlooked under pressure. Practicing these routines helps identify potential issues or shortages in your emergency supplies, providing an opportunity to address them proactively. These plans will be discussed in more depth throughout the book.

While each home and situation is unique, tailoring the staying-in strategy to fit specific needs increases its effectiveness. Assessing potential threats in advance and planning accordingly strengthens both physical and emotional defenses. Each aspect of this approach—security, resource management, and psychological preparedness—combines to establish a comprehensive, resilient plan for weathering various crises.

Given these advantages, staying inside should be considered the default strategy. During less volatile moments, prepare by stockpiling essential supplies and reinforcing home defenses. However, remain adaptable and ready to switch plans if the situation demands it. Flexibility is key. While staying in can be a primary strategy, circumstances such as extensive damage to the home or official evacuation orders may require a change in approach.

### Identifying Weak Points in Doors and Windows

When it comes to securing your home, a critical aspect is recognizing and addressing the physical weaknesses in its entry points, such as doors and windows. These are often the first targets for intruders, making their fortification essential in maintaining a safe environment. Here is how you can identify the weak points in your key access locations:

- **Inspect door frames:** Start by inspecting door frames for structural integrity, an inspection that might reveal signs of wear and tear that could be easily overlooked. Damaged or decaying frames can significantly compromise security, offering easier access to unwanted visitors. Pay particular attention to areas where frames join walls, as these junctures are common weak spots prone to deterioration over time. In addition, ensuring that the materials used in your doors are both durable and high-quality is vital; solid wood or metal doors are preferred choices due to their durability compared to hollow-core alternatives.
- **Check the windows:** Moving onto your windows, analyze them for potential vulnerabilities. The type of lock you use can make a significant difference in security. Basic window latches may not suffice against determined intruders, so consider upgrading to more secure locking mechanisms like keyed locks or smart locks that offer enhanced protection. We'll discuss better lock mechanisms throughout this book. The strength of the glass also plays a crucial role. Traditional glass windows can be easily shattered; thus, upgrading to laminated or tempered glass can provide added resilience against break-ins. These types of glass are designed to withstand impacts, deterring entry through forceful means.
- **Reinforce the frameworks:** In addition to strengthening existing frameworks, explore reinforcement options that can significantly enhance safety. Options such as security films, thick laminated glass, or plastic can be applied to windows to hold the glass together even if it gets broken. This makes breaking your windows more difficult and increases the time required for any intruder trying to gain access. Another effective measure is installing grilles or security bars on windows, which act as a physical barrier while still allowing natural light and air to flow through.

However, if considering this option, ensure they are installed with a quick-release mechanism for emergency exits during situations like fires.
- **Practice regular maintenance:** Routine maintenance and inspections ensure long-term security. Like any other household system, doors and windows require regular checks to maintain optimal condition. Look out for corrosion, particularly on metal components, and ensure all moving parts function smoothly without resistance, as stiffness or irregular movement might indicate underlying issues. Regular lubrication of hinges and locks can prolong their lifespan and effectiveness.
- **Consider seasonal changes:** Remember that seasonal changes can impact the integrity of these access points. For instance, wooden doors and frames may swell or shrink based on humidity levels, potentially affecting their fit and operation. Routine seasonal inspections can preemptively catch these issues before they become serious security concerns.

With all these factors in mind, don't forget to conduct regular assessments to catch early signs of weakness that might otherwise go unnoticed. For example, discovering small cracks in windowpanes or slight misalignments indoors can allow timely interventions such as repairs or replacements. Taking proactive steps in identifying and fixing these minor issues can prevent larger problems and costly fixes down the line.

## Assessing Electrical and Utility Vulnerabilities

One aspect that is often overlooked when protecting your home against vulnerabilities is the role of electrical systems and utility access. These components are crucial in maintaining a secure environment, especially during crises when power outages might occur. Understanding and addressing potential weaknesses can help ensure your family's safety and minimize unexpected disruptions. Some factors to consider when assessing these potential vulnerabilities include the following:

- **Monitor existing electrical and utility systems:** Firstly, it's essential to focus on monitoring existing electrical and utility systems. You should be aware of your home's electrical layout, including circuit breakers, wiring, and any accessible power lines. Regularly checking these systems for signs of wear or unauthorized access is vital. For instance, ensure that all meters and panels are securely locked and inspect for any tampering or unusual changes in electricity usage. Sometimes, intruders or unauthorized individuals may attempt to tap into your power supply unnoticed, leading to potential safety hazards. Continuously monitoring your electric infrastructure not only helps spot unauthorized access but also assists in the early detection of faults that could lead to larger outages. Note that implementing a

- **Use backup systems:** Secondly, considering backup systems such as generators or battery backups plays a significant role in safeguarding your home during power outages. A sudden loss of power can increase vulnerability, affecting everything from security alarms to emergency lighting. Backup power sources ensure those critical systems remain functional even when the main power supply fails. When selecting a generator or backup system, it's imperative to understand your home's specific needs. Calculate the energy requirements for essential systems during an outage and choose a generator that can accommodate them without compromise. Moreover, generators should be maintained regularly to guarantee their reliability when needed most. Regular testing, oil changes, and load assessments are fundamental practices to ensure backup units are ready for action.
- **Install safety devices:** Next, installing safety devices like surge protectors is another precautionary measure to shield your home from electrical vulnerabilities. Power surges are often unpredictable, resulting from lightning strikes, faulty appliances, or sudden power restorations. These surges can damage or destroy electronic devices and even pose fire risks. Surge protectors act as a defense line, absorbing excess voltage and preventing it from harming your home's infrastructure. Choosing the right surge protection devices involves evaluating their capacity and effectiveness. Look for products that offer ample wattage ratings and come with certifications ensuring compliance with safety standards. Place surge protectors strategically across the home, prioritizing areas housing valuable electronics or critical equipment tied to your security systems.
- **Incorporate professional regulation:** Finally, engaging professionals for regular inspections cannot be overstated. While it might be tempting to tackle minor issues independently, professional electricians bring a wealth of knowledge and experience that can uncover hidden problems invisible to the untrained eye. They can perform comprehensive analyses to ensure that all systems adhere to national safety regulations and guidelines, such as those outlined by OSHA or equivalent bodies. According to (*Electric Power Generation, Transmission, and Distribution*, n.d.), understanding the condition of protective grounds and equipment grounding conductors is essential in mitigating risks associated with induced voltages and other hazards. Scheduling biannual or annual check-ups with qualified professionals ensures peace of mind and complements your proactive measures in managing home security efficiently. Besides inspections, professionals can provide invaluable advice on potential upgrades or new installations necessary for enhanced performance and security.

You can consider using smart devices along with professional help to monitor any changes in your electrical systems and hardware. Smart meters and remote monitoring techniques can provide you with real-time data that can help you strengthen your security strategies.

Summary and Reflections on Understanding Your Home's Vulnerabilities

Throughout this chapter, you've learned how to evaluate and address potential vulnerabilities within your home and its surroundings. By conducting a comprehensive safety audit, you can pinpoint weaknesses in entry points, such as doors and windows, while also assessing the often-overlooked security measures within garages and basements. Involving every member of the household in this process ensures a shared responsibility for creating a safer environment, allowing them to identify flaws in your security that you might miss. Expanding this evaluation outdoors, we also covered how landscaping features could both enhance and hinder home security. Addressing these areas is key to maintaining clear sightlines and preventing tripping hazards, which are essential for quick evacuations in emergencies.

Furthermore, we've explored various strategies to enhance your security systems, emphasizing the importance of staying up-to-date with technological advancements and utilizing natural barriers effectively. Additionally, assessing the risks associated with electrical and utility systems provides an added layer of preparedness, ensuring that backup plans are in place during power outages or other crises. These combined efforts fortify your home while also creating an atmosphere of confidence and readiness. As you continue to implement these measures, remember that regular evaluations will help you maintain a secure and resilient home environment. Once you have a better understanding of the weak points in your home, you can make the executive decisions to boost your security, which often starts with the enhancement of your interior and exterior defenses. This topic will be covered fully in the next chapter.

# 2

## SECURING THE INTERIOR AND EXTERIOR OF YOUR HOME

How can you safely secure the interior and exterior of your home to protect you from potential crises? With an increase in prioritized family safety and growing uncertainties globally, it's essential to secure your residence. All homeowners face unique challenges when safeguarding their properties, ranging from potential burglaries to natural disasters.

It's important to note that securing a home means more than just installing traditional locks; it's about creating a comprehensive defense system that covers every potential threat. Through this chapter, we'll discuss all the detailed methods you can explore to make your space safer both internally and externally.

By learning about strategies such as installing perimeter fencing and upgrading door and window locks, you'll discover both mechanical solutions like deadbolts and advanced technologies such as smart locks. The importance of strategic outdoor lighting placement is also discussed, emphasizing its role in deterring crime through effective illumination. In addition, we'll examine the integration of security cameras and alarms, showcasing how these systems work in tandem to provide real-time surveillance. By the end of this chapter, you will gain a holistic understanding of how to secure your homes effectively, combining traditional methods with modern innovations to achieve peace of mind.

### Installing Durable Perimeter Fencing

The fencing around your property is the first line of defense against potential threats. This makes it valuable to install strong and durable fencing. There are various factors you should consider when installing your perimeter fencing, such as the following:

- **Choice of material:** The material selection for your fencing is essential, as it directly impacts the durability and effectiveness of this barrier. For instance, choosing materials like metal or reinforced wood can provide superior resistance to forceful entry attempts compared to weaker materials. Metal fences, such as those made from steel or wrought iron, not only offer strength but are also resistant to weather conditions and require minimal maintenance. Reinforced wooden fences can also serve as a formidable deterrent while adding aesthetic value to your property. Both options ensure a physical barrier that can withstand various attempts of breach, providing peace of mind for homeowners.
- **Choose suitable designs:** When considering the design of your fencing, height is an essential factor. Taller fences naturally discourage intruders from attempting to climb over them. A fence height that exceeds the average person's reach, typically over 6 ft, can effectively prevent unauthorized access. However, it's important to check local regulations regarding maximum fence heights to ensure compliance with legal standards. Besides height, incorporating features like spikes or barbed wire at the top of the fence can further deter climbing. While these features might seem extreme, they act as visual deterrents, signaling that breaching such a barrier would pose significant challenges. Anti-climb enhancements also come in various designs, allowing you to select options that blend well with your home's aesthetics without compromising on security.
- **Practice regular maintenance:** Regular maintenance of your fence is another key aspect to consider. Over time, even the most durable fencing materials may succumb to wear and tear due to environmental factors, such as rain, wind, or exposure to harsh sunlight. Regular inspections help identify any damage or weak spots that could compromise the security of your perimeter. This includes checking for any rust on metal fences or rotting sections in wooden ones. Prompt repairs and applying protective coatings or treatments can extend the lifespan of your fencing, ensuring it remains a solid and reliable defense mechanism. You should also keep the area around your fence clear of debris and vegetation to reduce hiding spots for potential intruders.
- **Use modern technology:** Incorporating modern technology into your fencing can further enhance its efficacy. Security fences today can be fitted with advanced features like surveillance cameras, motion sensors, and alarms. These devices can detect attempted breaches and also provide immediate alerts to you or your chosen security services, enabling rapid responses. By integrating these technologies, your fencing becomes an active participant in your home security system rather than just a passive barrier. This proactive detection capability allows you to stay informed and take necessary actions swiftly.

Apart from all these practical factors to consider, you can also be mindful of the aesthetic appeal of your fence. With numerous design options available, you can choose materials and styles that align with your personal taste while maintaining high-security standards. Whether choosing a modern metal design or a traditional wood finish, your fence can complement the style and appeal of your home. Well-crafted fences add structure and depth to outdoor spaces, making them more attractive. In fact, investing in a quality fence can increase property value, as potential buyers often view effective security measures as desirable features when evaluating properties.

Upgrading Door and Window Locks

You may have good quality door and window locks, but what if you could take them a step further by upgrading them and making them more durable? This can further enhance your defense against threats. Here are some factors to consider when improving your lock systems:

- **Make use of variety:** First, consider the variety of lock types available, each designed to meet specific security needs. Deadbolts, for instance, are renowned for their reliable mechanical security, minimizing risks associated with electronic breaches. They extend a solid metal bolt into the door frame, providing substantial resistance against forced entry attempts. Their simplicity and effectiveness make them a popular choice for homeowners seeking cost-effective solutions with minimal ongoing maintenance.
- **Consider smart locks:** On the other hand, smart locks offer advanced security features like encryption and real-time alerts, enhancing overall home security. These locks bring convenience and innovative access control options, allowing you to have keyless entry through smartphone activation, PIN codes, or biometrics. Integration with home automation systems adds further layers of security by synchronizing locking and alarm management, transforming traditional security measures into comprehensive, intelligent systems. While deadbolts function as reliable standalone measures, the inclusion of smart locks can boost protection, especially in emergencies when quick access control is crucial.
- **Get professional installation:** Next, you should consider your locks being installed by professionals. A correctly installed lock functions optimally, reducing vulnerabilities that may arise from DIY errors. Professional locksmiths ensure the perfect alignment and fitting of locks, maximizing their resistance to tampering or technical faults. Importantly, a professionally installed lock signals to potential intruders that the homeowner prioritizes security, which acts as an effective deterrent.
- **Maintain your locks:** Routine checks and maintenance significantly improve the lifespan and reliability of your locks. Regular inspections help identify wear and tear

or potential failures early, preventing situations where locks could fail during critical moments. Simple measures such as lubricating moving parts and checking for signs of rust can keep mechanical locks in good working order, while smart locks may need periodic software updates to protect against cyber vulnerabilities. Routine maintenance preserves the physical integrity of the locks while ensuring they remain effective barriers against unauthorized access.

Understanding the full spectrum of security options and implementations ensures you can choose the most appropriate combination of locking mechanisms and additional protective measures. While deadbolts and smart locks each have unique strengths and limitations, combining technologies offers a balanced approach, catering to both traditional security needs and modern conveniences.

Strategic Outdoor Lighting Placement

Outdoor lighting can have a huge impact on the safety of your home. By strategically using motion-sensor lights and floodlights, you can transform your surroundings into well-lit sanctuaries that deter crime. Dark areas around a property can attract intruders, providing them with the cover of night to approach unnoticed. The following tips can help you make the most of your outdoor lighting:

- **Find the right type of light:** Choosing a light that is most suitable for your security is the first step. Motion-sensor lights instantly eliminate these shadows, startling trespassers and casting unwanted attention their way. Floodlights, with their wide beams, illuminate larger areas, making it difficult for anyone to move undetected. This shining light alarms intruders and signals to you and your home's residents that there may be unusual activity outside.
- **Place lights optimally:** The placement of lighting is just as important as its presence. Lights should focus on entry points such as doors and windows, which are common targets for unauthorized access. Paths leading to and from these points must also be illuminated thoroughly. Properly lit pathways not only guide guests safely to your door but also make any movement along them conspicuous. According to Tim Rader, the senior director of product development at ADT, placing motion sensor lights along high-traffic areas, particularly near accessible windows and doors, maximizes their effectiveness. It's essential that these lights are mounted at an appropriate height to prevent tampering while maintaining optimal range for detection (Ahrnsen, 2024).
- **Program your lighting systems:** Modern programmable lighting systems offer additional layers of security by simulating occupancy. These systems can be set to mimic the patterns of household activity, such as turning lights on and off at varied

times throughout the evening. This simulation creates the illusion of someone being home, even when the occupants are away, effectively reducing the risk of break-ins. Programming your lights to adjust randomly can add to this deception, enhancing your home's defensive posture against potential burglars.

- **Maintain your lighting systems:** As with all your other security systems and methods, it's important to maintain them to keep them in good condition. Regular checks help detect issues such as burnt-out bulbs or bugs in the system that may hinder performance. Therefore, you should clean your fixtures, removing dust or debris that could affect light output. Ensuring lights have the correct power source and replacing batteries if necessary keeps them operational during critical moments. When you use solar-powered options, you should position them to receive sufficient sunlight so that they can stay charged.
- **Choose suitable materials:** Choosing the right materials for outdoor lights ensures durability against weather conditions. Engineered plastics or metals that resist corrosion are recommended, as these materials withstand heat, cold, rain, and wind better. In regions prone to extreme weather, checking for an IP rating, which measures ingress protection against water and dust, provides insight into how weatherproof a device is. The higher the IP rating, the better the protection provided, enhancing the longevity of your lighting system and maintaining consistent functionality over time.

The role of motion sensors is about more than just illumination, as they act as silent sentinels guarding your property. They should be tested regularly to confirm their sensitivity range aligns with your security needs. Rural homes might benefit from longer detection ranges due to fewer streetlights, while urban settings may require adjustments to prevent frequent triggers by pedestrian traffic. Some advanced models allow customization of sensitivity settings and lighting schedules, maintaining vigilance according to your preferences.

Setting up Security Cameras and Alarms

To secure your space successfully with the use of technology, you should set up security cameras and alarms. The integration of surveillance systems can significantly safeguard both the interior and exterior of a residence. Surveillance cameras serve as an ever-watchful presence, providing real-time monitoring and valuable recorded evidence in the event of any unlawful activity. Here's how you can strategically place cameras and alarms in and outside your home for security:

- **Consider your home's unique requirements:** When it comes to selecting the appropriate type of camera, understanding the unique requirements of your living space is crucial. Outdoor cameras are designed to withstand the elements—rain, wind, and sunlight—and should be rugged enough to function properly in varying weather conditions. These cameras are typically equipped with features such as night vision and wide-angle lenses to provide detailed coverage of exterior areas, including driveways, gardens, and other vulnerable spots. Indoor cameras, on the other hand, focus on monitoring the activities within your home. They can be strategically placed in common areas like living rooms and hallways or near entry points such as windows and doors to keep an eye on movement and potential intruders. The versatility of indoor cameras also allows them to double as pet or baby monitors, enhancing safety and peace of mind.

- **Position your cameras strategically:** Proper positioning of cameras is important in ensuring that both entrances and high-risk areas are adequately covered while minimizing the risk of vandalism. Entrances, such as front and back doors, are frequent targets for burglars. By placing cameras at these critical points, homeowners can monitor any suspicious behavior before a break-in occurs. Additionally, it's important to position cameras high enough to be out of reach of potential vandals but at an angle that still captures clear footage of faces and activities. Avoiding blind spots is key, so taking the time to assess different vantage points and perhaps testing camera angles during installation can prove beneficial.

- **Use alarm systems:** Modern alarm systems often come integrated with surveillance capabilities and are connected to professional monitoring services that provide rapid response in emergency situations. Studies have shown that the presence of security alarms influences the decision-making process of burglars, causing many to abandon their plans if an alarm is detected (*How Effective Are Home Security Systems?*, n.d.). This preemptive measure not only protects your property but also helps prevent further crime by alerting authorities promptly.

- **Don't forget software updates:** The effectiveness of a surveillance system does not only depend on physical hardware; regular software updates and consistent footage reviews are essential components in maintaining optimal performance. Just as frequent servicing keeps a car running smoothly, updating your security system's software ensures it remains resilient against hacking attempts and continues to operate efficiently. Reviewing recorded footage periodically allows homeowners to verify that the cameras are capturing clear images and working properly. It can also help identify patterns of movement around the property, which may inform adjustments to the positioning of cameras for improved security. According to Jennifer Pattison Tuohy & Serena Lopez (2024), offering video evidence to insurance companies can expedite claims processes and even help reduce premiums by demonstrating vigilance and preventive measures.

- **Consider the use of artificial intelligence:** Advancements in artificial intelligence technology have enhanced the functionality of surveillance cameras, allowing them to recognize faces, distinguish between people, vehicles, and animals, and send real-time alerts to homeowners when unusual activity is detected. These smart systems can be programmed to filter out false positives, such as tree branches moving in the wind, and focus on genuine security threats, further refining the level of protection offered. The ability to receive notifications instantly on mobile devices ensures that homeowners remain informed of any events regardless of their location, providing an added layer of reassurance.

If you're concerned about your privacy, it is possible to mitigate risks by opting for local storage rather than cloud-based options, where possible, and using features that allow control over when and what the cameras record.

### Reinforcing Internal Doors With Deadbolts

Although your entry points may look secure, you may find that strengthening these doors and windows provides additional security for you and your loved ones. Remember that your entry points are often the most vulnerable parts of any residence, so it's valuable to enhance their security by installing deadbolts.

Deadbolts significantly reduce the risk of forced entry, adding an extra layer of protection beyond standard doorknob locks. Unlike traditional locks that might be easily picked or kicked open, deadbolts feature a solid metal bolt that extends deep into the doorframe when engaged. This makes it incredibly difficult for burglars to manipulate or force open without considerable effort and noise, which can alert you or your security systems. When choosing a deadbolt, consider options like single-cylinder deadbolts, which require a key from only the outside, or double-cylinder deadbolts, which need a key from both sides, offering heightened security at the potential cost of quick exit during emergencies (*8 Easy Ways*, n.d.).

However, even the best deadbolt locks are only as effective as the doorframes they secure. Ensuring that your doorframes are solidly constructed and in good condition is vital. Weak or damaged frames can compromise even the strongest locks, making it imperative to reinforce them. Installing strike plates with longer screws that penetrate deeply into the frame can make them more stable. These strike plates help distribute the force of any attempted kick-in across a wider area of the frame, improving its integrity. Additionally, reinforcing frames with metal plates adds another layer of defense, preventing splintering or breaking under pressure (*8 Easy Ways*, n.d.).

To further fortify your entry points, additional reinforcements such as door braces and heavy-duty strike plates can be invaluable. Door braces provide a physical barrier that reinforces the door against being opened, even if the lock is compromised. Meanwhile, heavy-duty strike plates, designed to withstand significant force, ensure the door remains securely attached to the frame upon impact. Choosing these additional security measures can create formidable resistance against unauthorized entry attempts, dramatically enhancing the overall safety of your home (Abdelhadi, 2024,).

As with any addition and adjustment you make to your home, it's important to practice regular maintenance. After strengthening your entry points, regular maintenance and monitoring are essential. Over time, your locks and frames may experience wear and tear, which may negatively impact their effectiveness. Conducting routine checks helps you to identify any signs of damage or necessary repairs early on. For instance, checking whether door hinges are tight and that deadbolts properly align when locked can prevent issues from escalating. Equally important, lubricant application to locks can help them maintain smooth operation and prolong their lifespan.

## Concluding Thoughts

This chapter has explored practical strategies to enhance the security of your home from both external and internal threats. By focusing on the perimeter fencing as a first line of defense, with reliable materials, you can optimize security. Don't forget to consider factors such as fence height, design features, and regular maintenance. Through advanced locking mechanisms and smart technologies, you can have even more control over your safety.

You want all your entrances (i.e., doors and windows) to be safeguarded. From traditional deadbolts to modern smart locks, each offers unique advantages in fortifying entry points. Complementary security measures, such as window bars and door reinforcements, work alongside these locks to form comprehensive barriers. To complete the home's security ecosystem, strategic outdoor lighting and surveillance systems play crucial roles. Motion-sensor lights and programmable systems deter intruders, while well-placed cameras and alarms ensure consistent vigilance and rapid response. Together, these elements create a fortified environment that offers peace of mind and enhanced safety for families and individuals alike. With your entry points secured using these tips, it's highly unlikely that any breaches will occur. However, it's always best to prepare for any case scenario. In the next chapter, we will uncover how you can create an emergency plan that can prepare you for any event.

# 3

# DEVELOPING EFFECTIVE EMERGENCY PLANS

You never know what might happen in life, so it's crucial to be prepared with emergency plans that can secure you in any situation. Developing effective emergency plans will keep your family safe during crises. Having a well-crafted plan can be the difference between chaos and order, confusion and clarity. The process of building these plans requires thoughtful consideration and the unique factors of your family. From understanding potential risks to choosing methods of communication, every element plays a vital role in creating an effective strategy that addresses the specific needs of all involved. As emergencies can arise unexpectedly, preparation becomes a proactive step toward safeguarding your loved ones, ensuring they are equipped to handle any situation confidently.

This chapter delves into the intricacies of constructing a comprehensive emergency plan made specifically for your family's circumstances. You will explore how different communication protocols can be established to maintain contact and disseminate critical information efficiently during an emergency. It also examines the significance of designating key roles and responsibilities within the family unit, as clear task allocation can prevent panic and facilitate focused action. Additionally, it discusses practical approaches to practicing drills and scenario responses, ensuring everyone knows their part without hesitation. By integrating modern technology like mobile apps for alerts and GPS tools, you'll learn how to leverage available resources to enhance preparedness.

## Establishing Communication Protocols

When an emergency occurs, you may find it challenging to communicate with each other effectively because everything is chaotic and overwhelming. Communication plays a pivotal role during emergencies. When crises arise, having good communication with each other allows information to flow smoothly, so you can act with readiness and unity as a family. Here are some tips that can enhance your communication in emergency situations:

- **Primary contact person:** Designating a primary contact person is often overlooked but immensely important. This chosen individual acts as the central hub for all information and instructions, effectively reducing confusion that can easily arise in high-stress situations. By appointing someone reliable and organized, you and your family can ensure that emergency updates are centralized. Picture a scenario where multiple messages create chaos due to conflicting information. A primary contact prevents this by verifying and disseminating accurate details. This approach also shields your family from misinformation that can exacerbate an already tense situation. This structure provides clarity and direction, essential aspects during any crisis.
- **Alternative communication methods:** In today's technologically driven world, alternative communication methods like social media have revolutionized how we stay connected. When traditional channels fail—such as telephone lines getting jammed or power outages disrupting regular service—social media becomes invaluable. Platforms like Facebook, X, and Instagram offer instant updates and can serve as alternative communication methods to relay important information. When living in moments of crisis, posting updates through these channels keeps everyone informed and reassured. These social media platforms help you reach out to broader networks beyond immediate family members, allowing extended family or friends to stay informed about your safety.
- **Make a communication tree:** Creating a communication tree is another effective method to streamline information flow. This hierarchical model delineates who contacts whom, thus averting the possibility of missed messages or duplicated efforts. Just like branches of a tree, each family member is assigned to update specific individuals if an emergency arises. For example, you might be tasked with notifying other relatives while your partner connects with neighbors. This organization prevents overlap and ensures that key details are relayed promptly and efficiently. During chaotic moments, knowing exactly who to contact alleviates stress and confirms that no one is left uninformed.
- **Conduct regular check-ins:** Equally significant is the habit of conducting regular check-ins among your family members. These gatherings, whether virtual or face-to-face, promote consistent engagement and inspire an attitude of preparedness. Regular interactions enable family members to voice concerns, share insights, and refine their emergency plans based on the most recent developments or lessons learned. It's similar to holding a routine fire drill at home; the more practiced you become, the more instinctual your responses will be in real-time emergencies. Familiarity with each other's needs and expectations ensures swift action without

hesitation, greatly enhancing the overall resilience of the family unit in unpredictable situations.
- **Pretest your messages and lines of communication:** It's worth noting that pretesting messages and refining the delivery is paramount. Before relying solely on any channel, verify its effectiveness under varied conditions. Test how messages appear across different devices and networks, considering factors such as language accessibility and readability. Deliver clear, concise instructions that emphasize actionable steps without ambiguity, avoiding overly technical jargon. Instead, adopt a tone both authoritative and empathetic, recognizing the emotional strain inherent in emergencies.

As you work on all of these critical communication skills and tips, don't forget to be adaptable. Each family is unique and requires specific approaches to suit their dynamics and circumstances. Evaluate your plan regularly and adjust it when new technologies or challenges arise. Flexibility is your greatest ally in shaping a resilient communication system ready to withstand diverse emergencies.

## Planning Escape Routes and Meeting Points

In the unlikely event of an emergency, it's vital to have planned out safe evacuation routes that you and your family can use. Creating this plan facilitates a swift and orderly escape and also significantly reduces the stress associated with emergencies. Consider the following tips to create optimal escape routes and meeting points:

- **Have multiple exit points:** The importance of identifying multiple exit points is invaluable. Whether it's a natural disaster or a domestic situation requiring rapid departure, having more than one exit strategy offers the flexibility needed when conditions change unexpectedly. Consider a scenario where a fire blocks the primary exit route; an alternative can be the difference between safety and danger. By planning these pathways in advance, you can handle unforeseen circumstances.
- **Establish meeting points:** Establishing predetermined meeting points both within and outside your neighborhood is another critical aspect of effective emergency planning. During the chaos of an unexpected event, it is easy for family members to become separated. Setting clear locations to regroup post-evacuation minimizes confusion and ensures everyone is accounted for as quickly as possible. These meeting spots should be well-known and accessible to all family members, taking into account potential obstacles that might arise during an emergency. For instance, consider selecting a nearby park as a primary point and a more distant location as a secondary option, catering to varying emergency scenarios.

- **Practice regular drills:** Regular drills can embed these plans into memory, transforming them from an abstract concept into instinctual action. Practicing evacuation routes under different conditions—for example, in the dark or with simulated obstacles—builds muscle memory and confidence. It's similar to how athletes train; rehearsals improve performance by making actions automatic rather than conscious decisions. Scheduling these drills consistently helps ensure that every family member knows their role and can execute their part with minimal hesitation, even under duress.
- **Assess your surroundings:** It's necessary to assess your home's external surroundings periodically. Various factors, such as construction or natural changes, can alter the viability of previously identified routes. Regular evaluations allow for adjustments to ensure paths remain safe and open. Consider environmental changes like seasonal flooding or roadworks, which might temporarily block exits or meeting areas. Being proactive about these assessments aids in maintaining up-to-date and reliable safety strategies, safeguarding against disruptions posed by ever-changing environments.
- **Stay informed:** Finally, staying informed about community resources and protocols can improve your efforts and keep you in the loop. Knowing local emergency contacts, understanding public alert systems, and being aware of designated community shelters can greatly aid personal planning. In addition, engaging with local emergency services for advice or accessing governmental guidelines can further strengthen family plans, aligning them with broader safety measures in place.

Whichever route you decide to take, ensure that you tailor your routine and guidelines to what suits your family and home. Doing this ensures that you develop escape plans that are seamless and perfectly suited to you and your loved ones.

Designating Roles and Responsibilities

As a part of your effective emergency plans, you should designate roles and responsibilities for everyone involved. Doing this promotes teamwork and minimizes any confusion. When each member understands their role, there is a greater chance of smooth coordination and effective action during emergencies. The following tips can help you designate roles and responsibilities during stressful occasions:

- **Give each person a role:** Identifying each individual's strengths is the starting point for role assignment. By recognizing what everyone excels at, you align roles with their skills, which boosts effectiveness and instills confidence. For instance, if someone in your family is proficient in first aid, they could be designated as the primary medical responder in a crisis. This ensures that tasks are handled by those best equipped for them and builds each member's confidence, knowing they are contributing in a meaningful way.

- **Outline responsibilities:** Once individual strengths are identified, defining detailed responsibilities is the next step. Clearly outlined responsibilities help reduce ambiguity and confusion, ensuring every family member knows precisely what they are accountable for. In practice, this might mean assigning one member the task of gathering emergency supplies while another contacts external agencies or neighbors. Establishing these roles ahead of time leads to more decisive actions when it matters most, ensuring critical tasks are not neglected or duplicated.
- **Implement role rotation:** While specialization is vital, flexibility is equally important. Regularly rotating roles can boost resilience by preventing overreliance on specific individuals and ensuring that every member can step into different roles if necessary. For example, having multiple family members capable of operating emergency equipment or handling communication guarantees that the absence of one person doesn't jeopardize the overall response.
- **Discuss your emergency contingency plans:** Discussing contingency plans is another essential element. Emergencies often require adaptability as situations can change unpredictably. Contingency planning involves preparing for scenarios where designated roles may need to shift. If a key family member is unavailable, another should seamlessly assume their duties without hesitation. These discussions encourage a mindset of adaptability among family members, allowing them to respond effectively even under unexpected circumstances.

The importance of these practices can be illustrated by real-world applications where clear role definitions enhance outcomes. Drawing inspiration from health care settings, multidisciplinary teams lead to improved patient care by utilizing diverse professional expertise (Rosen, 2019). Similarly, having a structured approach to family emergency planning can ensure that all facets of a crisis are adequately covered, from securing the home to ensuring continuous communication with external parties.

Practicing Drills and Scenario Responses

Once you've put together a thorough contingency plan, it's time to practice drills to ensure these emergency plans are effective and swift. These drills also help prepare each family member to play their role and respond swiftly during real emergencies. By rehearsing various scenarios, you can test the effectiveness of the plans against unpredictable situations, enabling you to make any necessary changes. These tips can help you practice these drills as efficiently as possible:

- **Make sure your drills are realistic:** Conducting realistic practice drills is the best way to prepare yourself for different scenarios. Families should aim to replicate potential emergency conditions as closely as possible. For instance, a fire drill might include blocking one exit to mimic an actual obstruction, forcing participants to think on their feet and find alternative escape routes. Incorporating elements like turning off lights during a power outage simulation can further enhance realism. These exercises highlight the gaps in current plans, facilitating adjustments that could prove life-saving.
- **Practice multiple unpredicted scenarios:** Drills should not be limited to a single scenario to truly encompass the unpredictability of emergencies. Different crises call for different responses; hence, varying the situations helps prepare you and your family for anything from natural disasters to man-made threats. For example, practicing evacuation for both a tornado and a flood can reveal unique challenges and necessitate varied actions. Families will benefit greatly from this approach as it builds a versatile skill set that is adaptable to any crisis.
- **Debrief after drills:** Equally important is the process of debriefing after each drill. This reflection period is critical for pinpointing what went well and identifying aspects that need refinement. Openly discussing experiences encourages learning and improvement. Family members should review steps taken, evaluate decisions made under pressure, and contemplate alternative strategies for next time. Continuous learning through debriefing ensures that each drill polishes the collective response-ability, strengthening readiness over time.
- **Give and receive feedback:** Feedback during these discussions plays a pivotal role in refining the emergency plan. Every family member's input is valuable, as individuals may perceive or handle situations differently. Encouraging open dialogue fosters a collaborative atmosphere where everyone feels heard. By considering diverse perspectives, families can tailor their strategies to address specific needs and preferences uniquely. Children, for example, might voice concerns about aspects of the plan that adults overlooked, leading to more comprehensive solutions. Obtaining feedback encourages ownership among all participants. When family members see their suggestions integrated into the plan, they are more likely to remember and adhere to it during an emergency. This sense of contribution enhances commitment to the overall safety strategy, ensuring that everyone effectively collaborates when it matters most.
- **Create and use effective guidelines:** Incorporating guidelines is essential for conducting effective drills. Here's an example guide that can help you implement productive practice sessions:
    - Begin by defining clear objectives for each drill to keep the focus sharp and purposeful.

- Familiarize participants with the roles assigned to them beforehand to facilitate smooth execution.
- Then, simulate a realistic scenario and run through the drill without interruptions, allowing roles to be tested under simulated stress. However, safely integrate minor unexpected elements to encourage adaptive problem-solving.

Remember to have discussions after completing your drill to reflect on how well it went and make adjustments where you see fit.

Incorporating Technology in Emergency Planning

An effective way to elevate your emergency plan is by integrating the use of technology. Leveraging mobile apps for emergency alerts can keep you and all your family members informed about crisis developments. These applications, often developed with cutting-edge technology, offer real-time updates on weather conditions, natural disasters, and other emergencies that may arise in your area. For example, applications such as FEMA and Red Cross Emergency Alerts are valuable tools designed to provide timely notifications. They can be customized to suit specific needs, allowing you to set preferences for the types of alerts you wish to receive. By integrating these apps into daily life, you can maintain continuous awareness of potential risks, thereby enhancing your preparedness.

Another key advantage of using mobile apps for alerts is their portability and the fact that smartphones are usually within reach. This constant access means you and your family can receive critical information anytime, no matter where you are. Furthermore, many of these apps allow for personalization, which means you can select only the notifications relevant to your geographical location or specific concerns, reducing false alarms and unnecessary panic.

If you're interested in developing a digital emergency plan, here are some factors you can consider:

- **Prepare essential documents:** Creating a digital emergency plan involves compiling essential documents and contact information in an easily accessible format. This step ensures that even during the chaos of an emergency, vital information is at your fingertips. Start by digitizing important documents, such as identification papers, medical records, insurance policies, and property deeds. Store these files securely in cloud storage services like Google Drive or Dropbox, granting access to trusted family members. Doing so will secure these documents against physical damage and ensure they're accessible from any location.
- **Include emergency contacts:** In addition to documents, a comprehensive digital plan should include a list of emergency contacts, including family members, doctors, and local emergency services. Make sure to update this list

regularly, reflecting any changes in contact information or new additions to your circle of trust. Many apps and devices now offer the ability to store this information directly within the device, providing quick access through widgets or emergency service menus. This feature is especially useful when time is of the essence, enabling immediate communication without having to search for contact details.

- **Make use of GPS tools:** Using GPS tools during emergencies can significantly improve your ability to navigate efficiently and safely. Applications like Google Maps or Apple Maps come built-in with real-time traffic updates and route suggestions. In an emergency, these tools can highlight optimal evacuation paths, taking into account road closures, traffic jams, and other obstructions. This allows you to make informed decisions, selecting routes that minimize risk and time spent exposed to danger. GPS-based applications like Waze also provide community-driven information, offering updates from fellow users on current road conditions. During crises, this collective intelligence can be invaluable, alerting users to hazards that might not yet be officially reported. Additionally, some emergency-focused apps integrate GPS technology to guide users toward designated shelters or safe zones, which is especially useful when navigating unfamiliar territories.

- **Use group chat platforms:** Establishing group chat platforms is another effective strategy for seamless communication during emergencies. Platforms like WhatsApp, Signal, or GroupMe help you share and receive information rapidly, keeping everyone in the loop with real-time updates. Through features such as group messaging and broadcast lists, families can coordinate plans, share locations, and provide status updates effortlessly. Setting up a dedicated family group chat for emergencies ensures that all members receive consistent messages simultaneously, reducing confusion and the need for multiple phone calls. These platforms often support multimedia sharing, enabling you to send pictures, videos, or voice notes to convey detailed information quickly. In high-stress situations where verbal communication may fail, such capabilities are indispensable.

- **Consider setbacks:** As powerful as these technological tools are, it is equally important to understand the limitations and build contingencies into the planning process. Always consider backup options, such as printed copies of the emergency plan and contacts, and have alternative communication methods ready in case your technology fails. Battery drains, network outages, or equipment failures shouldn't leave you unprepared. For instance, investing in portable chargers or power banks can ensure that devices remain operational during extended power outages. Likewise, knowing how to manually interpret maps or signals could provide essential skills if GPS services are unavailable. Moreover, establishing clear protocols for meeting points or designated locations where information boards or traditional landlines are available can serve as invaluable resources.

The integration of technology into emergency preparedness efforts demands ongoing adaptation and learning. As technological solutions continue evolving, staying updated with the latest advancements enhances the effectiveness of family emergency plans. Participating in online forums, workshops, or informational sessions related to emergency readiness can keep you informed about new tools, applications, and best practices.

## Concluding Thoughts

Crafting a comprehensive emergency plan specified to your family's needs involves several crucial components. Firstly, effective communication protocols ensure that every family member is kept informed and prepared during emergencies. By designating a reliable primary contact person, confusion can be minimized as they become the central point for information dissemination. In addition, alternative methods like social media can be valuable when traditional channels fail, offering avenues to maintain connection and relay critical updates. Don't forget to establish communication trees and conduct regular check-ins to create an effective network ready to adapt to any crisis situation.

Equally important are strategies focused on escape routes and meeting points, along with clearly defined roles and responsibilities. Planning multiple exit pathways and setting predetermined gathering spots within and outside the neighborhood help you respond swiftly and orderly during crises. Assigning roles based on each individual's strengths enhances coordination and boosts confidence, while role rotation ensures flexibility. Regular drills transform plans into instinctual actions, making them effective when seconds count. By incorporating technology, such as mobile apps for alerts and GPS tools, you can enhance your awareness and navigation capabilities. Combining these elements into a cohesive plan cultivates resilience, ensuring that your family remains safe and secure in the face of unpredictable emergencies. With thorough emergency plans in place, it's time to prepare your home with the resources you need to survive any event. In the next chapter, we'll be covering essential resources for you and your family.

## 4

# ASSEMBLING ESSENTIAL RESOURCES—FOOD, WATER, AND OFF-GRID ENERGY SOLUTIONS

During stressful emergency situations, the last thing that may be on your mind is the resources you need to continue surviving and thriving. Gathering and managing essential resources such as food, water, and off-grid energy solutions beforehand prepares you for these emergencies. In scenarios where everyday conveniences may become unavailable without warning, having a strategic plan in place can be critical for survival.

While the unpredictability of natural disasters or crises often leaves people feeling vulnerable, understanding and preparing for basic needs provides stability and empowers you and your family to face challenges with confidence. Exploring reliable methods to secure these necessities fosters a proactive approach toward self-reliance and safety, making it an invaluable pursuit for any household focused on preparedness.

This chapter discusses practical strategies for compiling robust reserves of food and water while also exploring sustainable off-grid energy solutions. You will discover effective budgeting techniques for stocking up on long-lasting staples like wheat, corn, and beans, alongside selecting emergency foods known for their durability and nutritional content. The discussion extends to water management plans that emphasize filtration systems and purification tablets for safeguarding against contaminants. You'll also discover information about organizing and rotating supplies to help maintain freshness and reduce waste.

Beyond food and water, this chapter highlights the importance of adopting renewable energy sources such as solar panels and wind turbines, supported by efficient storage solutions like batteries. These strategies are further enhanced through the integration of modern technology, ensuring both resource efficiency and household sustainability.

Compiling Food and Water Reserves

During times of crisis, having enough food and water is crucial for your health and survival. To prepare for these unforeseen disasters or emergencies, it's valuable to compile adequate reserves of food and water.

The concept of maintaining well-organized food and water supplies is not merely about survival; it's about empowering you to face any situation with preparedness and assurance. The following steps can help you put together a well-rounded reserve filled with all the food and drinks you require to thrive:

- **Create a structured budget:** The first step toward building a reliable emergency food supply involves creating a well-structured budget. A focus on bulk purchases helps save money and build sufficient stocks that are available when needed. Buying in bulk is often more cost-effective because suppliers typically offer discounts for larger quantities. This purchasing method allows you to maintain an ample supply of staple foods without frequent trips to the store, which may not be possible during emergencies. Bulk staples like wheat, corn, and beans, proven to have long shelf lives, are prudent choices for stocking up. These items are nonperishable and can last for years if stored correctly, making them ideal candidates for long-term storage plans. Creating a budget with consideration for bulk purchases establishes a solid financial foundation for sustaining readiness over time.
- **Select appropriate foods:** When it comes to selecting the right types of emergency foods, focus should be placed on options known for their durability and nutritional content. Canned goods, freeze-dried meals, and Meals Ready-to-Eat (MREs) stand out as excellent choices due to their extended shelf lives and ease of preparation. Canned goods are resilient against spoilage, while freeze-dried and dehydrated options provide the added benefit of lightweight and compact storage. MREs, initially designed for military use, are complete meals requiring minimal preparation, which could prove invaluable in situations where cooking facilities are limited or unavailable. Prioritizing these versatile selections ensures that nutrition remains intact over extended periods, reducing concerns about the deterioration of food quality.
- **Manage your water:** Equally vital to food storage is water management. You should store at least 1 gal of water per person per day, factoring in both hydration and food preparation needs. Having an established water storage plan means securing an adequate amount to cover essentials for drinking, cooking, and hygiene. It's advisable to invest in proper filtration systems or purification tablets, ensuring that stored water remains safe for consumption. We'll discuss some effective purification methods shortly.
- **Create rotation systems:** To maximize the effectiveness of food and water

reserves, a rotation system should be implemented. This process involves organizing supplies so that older items are used first, thereby reducing waste and preventing expiration. Marking dates of purchase and usage on packages aids in tracking freshness and ensuring rotation occurs efficiently. Regularly rotating stock means that foods are consumed well before expiration, maintaining their nutritional value and flavor. This practice preserves the quality of your supplies while promoting household efficiency by minimizing unnecessary disposal of expired goods.

Incorporating these practices into daily routines ensures that you and your family are not scrambling for solutions during potential panic. Should disaster strike, the strength drawn from preparation allows you to focus on other critical survival tasks instead of worrying about basic sustenance. A proactive mindset coupled with practical strategies secures your physical necessities and provides mental reassurance during stressful events. When you focus on storing your basic essentials, preparedness can become second nature for you, equipping you with skills to approach emergencies confidently.

## Food Stockpiling & Storage

Food stockpiling and storing your essential foods take the pressure and stress off events where you may be in need of your basic necessities. To stockpile and store your foods successfully, consider the following tips:

- **Understand your foods:** It's essential to recognize that not all foods are created equal when it comes to shelf life and nutritional value. Foods with extended shelf lives, like rice, beans, and canned goods, are ideal for stockpiling since they maintain their quality over time and provide necessary nutrients. Understanding these aspects allows individuals to make informed decisions about which items will best sustain their household during a crisis.
- **Identify nutritional needs:** Delving into nutritional needs is equally crucial. In times of emergency, maintaining good health requires consuming well-balanced meals that cover essential dietary requirements. A typical stockpile should include proteins, carbohydrates, fats, vitamins, and minerals. For instance, beans offer plant-based protein, while rice provides energy-boosting carbohydrates. Incorporating a variety of food items ensures diversifying meals and meeting your body's nutritional needs even when grocery stores might be inaccessible.
- **Consider temperature control:** Ideally, store perishable goods in a cool, dry area. Basements or cellars often provide optimal conditions, but ensure proper insulation to prevent temperature fluctuations. Note that the stability of the environment enhances the longevity of stored foods and assures readiness when they are needed most.

- **Organize your stockpile:** Efficiently organizing your stockpile ensures easy access and monitoring. Labeling each container with the name of the item and its expiration date simplifies tracking inventory and promoting a first-in, first-out system. This approach minimizes waste, enabling families to use products before they expire. Categorizing items based on meal type or nutritional content further enhances this system, facilitating meal planning and preparation when time is constrained.
- **Buy wisely:** Consideration of dietary restrictions cannot be overlooked. Inclusivity in stockpiling practices ensures that every family member's unique dietary needs are met, fostering well-being and unity during challenging times. Whether someone has allergies or follows a specific diet, accounting for these requirements strengthens the resilience of the household. By having suitable meal options available, no one is left without adequate sustenance. Lastly, it's important to include foods you enjoy. Stockpiling meals your family regularly consumes not only maintains morale but also guarantees efficient usage. Familiar foods bring comfort amidst uncertainty, helping ease the psychological toll of adapting to emergency life.
- **Creating your first stockpile may feel overwhelming.** Buying items with a long shelf life is a practical starting point, particularly with inexpensive, easily storable staples like rice and beans (Thomas, 2020). These foods require minimal preservation effort and can serve as the foundation for versatile meals. Moreover, having ready-to-eat options, such as canned soups or instant noodles, adds convenience since they are especially useful when cooking facilities are unavailable or quick meals are required. As you grow your food storage, you will learn what works for you and what doesn't.

## *Essential Foods to Store for Emergencies*

We've discussed some effective foods to store in your stockpile, but let's take a more in-depth look into foods you should store for unexpected emergencies. These foods should be easy to prepare and nutritious and have a long shelf life.

### Canned Foods

Canned foods are a staple for any emergency food supply. They come in various forms, such as vegetables, fruits, meats, and soups. Canned vegetables and fruits can provide essential vitamins and minerals. Look for low-sodium options for vegetables to keep the salt intake in check. Canned meats like chicken, tuna, and salmon are excellent protein sources and can be used in many recipes. For example, you can mix canned tuna with some mayonnaise and relish for a quick, easy meal.

Be sure to check the labels and select items with longer expiration dates. Always store them in a cool, dry place to maximize their shelf life.

## Grains

Grains are another important category of food to include in your emergency stockpile. Items like rice, pasta, oats, and flour are versatile and can be used in many different dishes. For rice, you might choose quick-cooking varieties for easier preparation. Pasta can be boiled and served with a simple sauce, while oats can be turned into a hearty breakfast.

Whole grains are generally more nutrient-dense than refined grains. They contain more fiber, which is beneficial for digestion. You can also consider storing quinoa, which is a complete protein that cooks relatively quickly.

## Legumes

Legumes, such as beans and lentils, are affordable and nutrient-rich. They are excellent sources of protein and fiber and can be very filling. Dried beans can be stored for years but may require soaking and cooking, while canned beans are ready to use immediately.

Consider using black beans in a soup or lentils in a stew. Both options are easy to prepare and can be flavored with various herbs and spices to elevate the meals. Legumes can also be made into dips, such as hummus, by blending with tahini and lemon juice.

## Nut Butters

Nut butters like peanut butter and almond butter are great additions to your emergency food supply. They are high in protein and healthy fats, which can help you feel satisfied. Nut butters can be eaten straight out of the jar or spread on crackers or bread.

When selecting nut butters, look for options without added sugars or hydrogenated oils for a healthier choice. They can also be added to smoothies or oatmeal for extra flavor and nutrition.

## Freeze-Dried Foods

Freeze-dried foods are a great option for long-term storage. They retain most of their nutrients and can last many years when properly stored. Companies offer freeze-dried meals, vegetables, and fruits that can be rehydrated with water.

You might find freeze-dried fruit a tasty snack or topping for yogurt. Freeze-dried meals can be a convenient option, especially during emergencies when cooking might not be possible. Just be sure to follow the rehydrating instructions for best results.

## Snacks

While practical food items are essential, including snacks is also important for maintaining morale during tough times. Items like granola bars, trail mixes, and crackers are portable and easy to store. Granola bars can provide a quick energy boost when you need it, while trail mix supplies a mix of nutrients from nuts, seeds, and dried fruits.

When choosing snacks, consider those that have longer shelf lives. Look for options with lower sugar content for a healthier choice. Snacks can be comforting and help you feel more at ease during emergencies.

**Cooking Essentials**

Having cooking essentials on hand can make meal preparation easier during emergencies. Items such as cooking oil, salt, and various spices can enhance the flavors of your meals. For example, a simple dish of rice and beans can be made more enjoyable with just a sprinkle of salt and pepper.

Another essential is sugar, which can be used for sweetening foods or beverages. You might also want to consider baking powder and baking soda, as they can be useful in preparing a variety of baked goods, even in stressful situations.

**Beverages**

Don't forget about beverages! Storing adequate water is crucial during emergencies. Aim for at least 1 gal of water per person per day for drinking and sanitary uses. Additionally, consider storing shelf-stable beverages like powdered milk or juice boxes.

Powdered milk provides a good source of calcium and can be reconstituted with water for drinking or cooking. Moreover, filling your pantry with tea and coffee can provide comfort and routine during difficult times.

**Vitamins and Supplements**

Maintaining health during emergencies is vital, so consider storing vitamins or supplements. They can help fill nutritional gaps, especially if your food supply does not cover all food groups. Multivitamins can provide a broad range of nutrients that can support overall health.

While not a replacement for actual food, these supplements can be a useful addition to your emergency preparedness plan, ensuring that your family gets some essential vitamins and minerals, even when fresh produce is not available.

Planning and assembling a solid emergency food supply involves careful consideration of a variety of food groups. By thoughtfully selecting foods that are nutrient-dense, long-lasting, and easy to prepare, you can ensure that you're equipped to handle emergencies with confidence.

Water Sources & Purification

Nobody can live without water. When emergency situations occur, it's important to guarantee that you and your family stay fully hydrated. In these scenarios, you should ensure you have access to clean and reliable water. The following strategies can supply you with water when you need it most:

- **Identify water sources:** Identifying nearby natural water sources, such as rivers, lakes, or streams, is often the first step in creating a plan for water acquisition. These natural sources can provide a significant advantage during shortages when typical supply lines are unavailable. However, it's crucial to approach these options with caution and purification strategies as they may be contaminated with pathogens.
- **Purify water:** Once a water source is identified, effective purification methods must be used to ensure the water is safe for drinking. Boiling is one of the most reliable methods, as it kills harmful microorganisms. Bringing water to a rolling boil for at least 10 minutes effectively eliminates the vast majority of bacteria and viruses. This method is particularly beneficial because it requires no special equipment beyond a heat source and a fireproof container.
- **Filter water:** Another common purification method involves filtering. Portable water filters, which are lightweight and easy to use, can remove bacteria and other contaminants from water. Many filters are designed to handle small particles, making them an essential tool for hikers, campers, and anyone preparing for emergencies. If you want to be ready for any possible situation, including chemical treatment options like iodine tablets or chlorine drops in your preparedness kits can make a big difference. Even though these chemicals need precise dosage control, they are convenient solutions for killing pathogens.
- **DIY water purification:** Do-it-yourself (DIY) water purification systems offer another layer of security, especially when commercial products aren't available. These systems can be constructed using basic materials found at home or in nature. A simple DIY filter might include layers of sand, charcoal, and gravel housed in a plastic bottle or a similar container. While these homemade filters may not entirely purify the water, they effectively remove larger debris and sediment, serving as a preliminary step before boiling or chemically treating the water. Through experimentation and practice, you can learn how to assemble and utilize these systems efficiently, boosting your self-reliance in times of need. You can find a DIY purification method at the end of this section.
- **Maintain your water's quality:** Maintaining stored water's quality and safety is necessary to ensure it remains potable over time. Regular testing of stored water is recommended to detect any changes in quality. Simple water test kits can be used to check for bacterial contamination, and their usage should form part of a routine maintenance schedule. Keeping track of expiration dates on water treatment supplies and regularly rotating the stock ensures freshness and readiness. Additionally, appropriate storage conditions, such as keeping water in cool, dark places and shielded from direct sunlight, help prolong its shelf life.

- **Regularly review stored water:** If you rely heavily on stored water, implementing a regular review process to ensure supplies remain uncontaminated is essential. This proactive approach provides peace of mind and prepares you for uncertainties ahead. Moreover, maintaining a balance between stored water and plans for acquiring new sources ensures the long-term sustainability of water resources in any emergency scenario.

When you prepare yourself and your home with essentials, don't forget your water! By understanding the location of viable water sources, utilizing diverse purification techniques, and regularly checking stored water quality, you can guarantee your and your family's well-being during crisis situations.

### *Creating a DIY Water Purification System*

When you want to ensure you have access to clean water, building a DIY water purification system can be both useful and rewarding. This process can be done with a few simple materials and tools.

**Gathering Materials**

The first step in creating this system is to gather your supplies. You will need a plastic or glass container, sand, activated charcoal, gravel, and coffee filters or a clean cloth. Each of these materials plays a significant role in the purification process.

1. A container will serve as the base for your purification system. A large plastic bottle can work well.
2. Next, you will need sand. Sand is an important part of the filtration system because it can trap dirt and debris.
3. Activated charcoal is also needed; this material helps eliminate odors and harmful chemicals.
4. Gravel will act as another layer of filtration.
5. Lastly, coffee filters or a clean cloth will be used to cover the top of the container.

**Setting up Your System**

Once you have gathered all your materials, it is time to set up your purification system. Start by taking your container and turning it upside down. This method allows the filtered water to flow out from the bottom. The first layer you will add is gravel. Pour a couple of inches of gravel into the bottom of the container. This layer helps filter out larger particles.

After adding gravel, the next layer should be sand. Add a thick layer of sand on top of the gravel. The sand will help trap smaller particles and contaminants that the gravel may miss. It is essential to make sure that the sand is clean to ensure effective filtration.

Following the sand, you will add the activated charcoal. This layer is crucial as it removes impurities and improves the taste of the water. Spread a layer of activated charcoal several inches thick.

The final filter layer is where you will use the coffee filter or clean cloth. Place it on top of the activated charcoal. This filter prevents any fine particles from escaping into the purified water.

**Filtering Water**

Once your setup is complete, the next step is to start filtering water. Begin by collecting the water you want to purify. It could be from a river, a pond, or any other source that is not immediately safe to drink. It is important to remove any large debris from the water beforehand to prevent clogging your filter.

Pour the collected water slowly into the top of your container. As water passes through each layer of gravel, sand, and charcoal, it will be filtered. The time it takes for the water to filter through can vary depending on how much you pour in. You may need to wait a few minutes for the water to completely filter through the layers.

**Testing for Cleanliness**

After the water has passed through your filter, it is advisable to test it for cleanliness. There are various ways to do this. One method is to use test strips that check for bacteria and other impurities. If you do not have access to test strips, you can boil the water, which can help kill any remaining pathogens. Aim to boil the water for at least one minute or longer if you are at high altitude.

**Final Steps**

Once you are confident that your water is safe to drink, you can start using it. It is a good practice to keep your DIY water purification system clean. Regularly clean the container and replace the filter materials as needed. This will help ensure your filtration system remains effective.

Also, if you find yourself using this system frequently, consider setting up multiple filters. This way, you can purify larger quantities of water more efficiently. Have an extra container ready so you can filter water simultaneously. This makes the process faster and guarantees you always have access to clean water.

**Maintenance and Care**

Keeping your water purification system in good condition is vital. After each use, make sure to clean out any sediment that may have settled at the bottom. Rinse out the container with clean water and let it dry. It is also important to replace the activated charcoal and sand regularly, as these materials can become saturated over time.

By maintaining your DIY water purification system, you can ensure a continuous supply of clean water. This process empowers you to take control of your water quality, especially in outdoor adventures or during emergencies.

Creating a DIY water purification system is not only practical but also a valuable skill to have. Whether you are in the wilderness or just preparing for unexpected situations, knowing how to make clean water can make a significant difference. Following these guidelines can lead you to a reliable solution for your clean water needs.

## Off-Grid Energy Solutions

Being off-grid means you never have to worry about electricity disappointing you and your emergency procedures. Fortunately, we have a variety of energy solutions to choose from, so it's valuable to find an option that works best for you. Here are some energy solutions to consider:

- **Solar panels:** Solar panels have emerged as a prominent aspect of sustainable off-grid energy solutions. By harnessing sunlight, they offer a clean and renewable power source, significantly reducing dependence on traditional electricity grids. This transition minimizes carbon footprints and empowers households to maintain energy autonomy. Solar panels are relatively simple to install and can be tailored to fit various needs and budgets through scalable setups, ranging from small residential systems to larger arrays for more extensive power requirements.
- **Wind turbines:** For regions blessed with consistent wind patterns, wind turbines present an innovative supplement to solar energy. These turbines capture kinetic energy from the wind, converting it into electrical energy. They are particularly effective in coastal areas, open plains, or mountainous regions where wind speeds are optimal. Installing small-scale wind turbines can diversify the energy mix and improve overall sustainability by ensuring power availability even during cloudy or sunless days. The combination of solar and wind energy can create a hybrid system that maximizes energy generation across different weather conditions, offering a holistic solution to meet continuous energy demands.
- **Generators:** Generators remain a vital component in the toolkit of emergency preparedness, offering immediate and reliable backup power. They provide a critical buffer during unforeseen outages when neither solar nor wind energy is available, ensuring that essential appliances and systems continue to function. Modern generators, especially those powered by cleaner fuels like propane or natural gas, come equipped with advanced features for efficiency and reduced emissions. On the other hand, portable generators are particularly beneficial in emergencies, allowing families to access power swiftly without complex setup procedures. It's essential to

regularly test generators and keep them well-maintained to guarantee their readiness in urgent situations.

## *Understanding Energy Usage*

Being mindful of energy storage is fundamental in optimizing resource efficiency within an off-grid setup.

### **Batteries**

Batteries serve as the backbone of energy storage, capturing excess electricity generated by solar panels or wind turbines for later use. Advances in battery technology have led to more compact, efficient, and durable solutions. Lithium-ion batteries are prevalent due to their longevity and high storage capacity, making them ideal for residential use. Beyond choosing the right type of battery, it's crucial to implement a smart management system that monitors and regulates energy flow, helping to maximize the lifespan of both the storage units and the renewable energy system as a whole.

### **Energy Audit**

Adopting these strategies requires careful consideration and planning. Conducting an energy audit is a recommended first step to understanding a household's specific consumption patterns and identifying opportunities for efficiency improvements. This analysis helps configure the most appropriate mix of solar panels, wind turbines, and generators tailored to the unique environmental conditions and energy needs of the home. In areas with less predictable weather, incorporating energy-efficient appliances can reduce demand and enhance reliability.

### **Smart Technology**

Integrating smart technology into these systems enhances control and monitoring capabilities. Smart inverters, for example, optimize energy conversion and distribution, while remote monitoring solutions allow for real-time tracking of energy production and consumption. Such technologies provide homeowners with valuable insights into their energy behavior, enabling proactive adjustments to usage patterns that conserve resources and save costs.

### **Sustainability and Maintenance**

Sustaining off-grid energy systems involves regular maintenance checks and updates. Solar panels and wind turbines require occasional cleaning and inspection to maintain peak performance. Ensuring connections are secure and free of obstructions is vital for efficiency. Likewise, generator maintenance should include checking fuel levels, oil changes, and servicing mechanical components. Establishing a routine maintenance schedule can prevent breakdowns and extend the operational life of the equipment.

Communication Devices and Power Sources

It's valuable to explore different types of communication devices and power sources that ensure you always have a reliable way to stay connected with your family. This involves more than just keeping the power on; it means prepping yourself with the right tools and strategies for effective communication and energy management. Here are some must-haves and valuable strategies that can help you keep communication alive even during emergencies:

- **Keep mobile phones on hand:** Mobile phones remain one of the most essential communication devices in emergency scenarios. Their ability to access multiple networks allows for versatility in communication methods. Text messages and emails, which operate independently of clogged voice lines, can often reach their destinations even when other methods fail (*Top 10 Communication Methods*, n.d.). Having a landline telephone can be a wise investment, as it might still function when cell service is disrupted due to its reliance on a different technology infrastructure.
- **Use two-way radios:** Two-way radios offer another layer of reliability, particularly when mobile networks become overloaded or fail entirely. These devices, also known as walkie-talkies, enable direct person-to-person communication over short distances without relying on cellular infrastructure. They are invaluable for coordinating efforts among family members or responders in real-time situations. More advanced versions like Citizen's Band (CB) radio and ham radio provide an expanded range and greater functionality but require licensing and knowledge to operate effectively (*Top 10 Communication Methods*, n.d.).
- **Integrate emergency alert systems:** In addition to personal devices, integrating emergency alert systems into your preparedness plan ensures you receive timely information about weather conditions or civil disturbances. Systems such as the Google Crisis Response platform offer a variety of tools and interactive platforms that aid both individuals and emergency responders during disasters (*Top 10 Communication Methods*, n.d.). Notifications from these alerts can guide decision-making processes, such as evacuation timing or sheltering-in-place procedures.
- **Have a communication plan:** Developing an effective communication plan is another crucial step in ensuring cohesive and efficient responses during emergencies. This plan should include designated meeting points where family members can reconvene if separated, as well as predetermined contacts who can assist in relaying information when direct communication is unavailable. Having clear instructions about who to contact and when streamlines efforts and reduces confusion during chaotic times. Encourage family members to memorize critical phone numbers or keep a written list readily accessible in case digital devices fail.

As with previously mentioned security tips, don't forget to find backup options for when your power may be out. Additionally, use modern technology to offer more safety and connectivity. Apps like Life360 and FEMA provide location-based services and emergency checklists, helping you and your family track each other's whereabouts and access lifesaving resources during critical moments (*Protective Actions Research*, 2024).

Concluding Thoughts

Managing critical resources like food, water, and energy forms the foundation of effective emergency preparedness. This chapter has explored practical strategies for homeowners, renters, and families to ensure readiness during crises. From establishing sufficient food and water reserves using bulk purchasing and rotation systems to understanding off-grid energy solutions like solar panels and generators, you can approach the unexpected events of life with confidence and clarity. By adopting these methods, your family can survive any scenario, and you're able to maintain a sense of normalcy in adverse situations. The emphasis on creating organized stockpiles and implementing efficient storage practices underscores the importance of being proactive.

We've also addressed effective communication during emergencies as an important strategy using essential tools and systems to keep families connected and informed. Devices ranging from mobile phones to two-way radios, coupled with reliable power sources, ensure that lines of communication remain open when they are needed most. Through comprehensive planning, including setting up communication plans and leveraging modern applications, families can navigate crises more seamlessly. Preparing with these strategies allows you to develop resilience and security, enabling you to protect your loved ones with certainty and readiness. Being prepared in as many ways as possible is paramount. Thus, in the next chapter, we will discuss how you can achieve medical preparedness.

# 5

# MEDICAL PREPAREDNESS—ADDRESSING MEDICAL EMERGENCIES AND CARE

Being medically prepared for any scenario can keep you safe during any unexpected emergency. It's important to realize that unforeseen situations can arise at any moment, which makes it invaluable to respond promptly and effectively. Through this chapter, we will uncover the various components that encourage comprehensive medical preparedness, addressing both the physical supplies needed and the essential knowledge you should possess. By exploring these factors, you can grasp the importance of readiness in protecting yourself or your loved ones from potentially life-threatening scenarios. With numerous factors contributing to your preparedness level, understanding how to integrate each element seamlessly into daily life can significantly enhance personal and family security.

This chapter will guide you through several key areas necessary for effective medical preparedness. It'll cover the basics of assembling a well-rounded first aid kit tailored to individual and family needs, emphasizing the importance of adaptability in emergency supplies. It'll also uncover the role of training in essential first aid techniques, such as CPR and appropriate use of first aid materials, empowering you with the skills to confidently respond when faced with emergencies.

The integration of prescription and over-the-counter medications within your preparations is also discussed, highlighting the benefits of natural remedies and self-care practices as supplementary strategies. Furthermore, the chapter addresses special considerations for families with unique medical needs, ensuring that all members have access to the appropriate resources and support. Ultimately, this comprehensive examination provides you with the knowledge required to create an effective plan for medical readiness during diverse and unexpected challenges.

## First Aid and Medical Supply Considerations

Anything can happen during an emergency, so having the right first aid and medical equipment handy can make all the difference. In the event of unexpected injury or illness, having immediate access to necessary supplies is critical. Here are the considerations you should make when putting together your emergency first aid kit:

- **Add your basic essentials:** To start any successful first aid kit, you should include basic items like bandages, antiseptics, and pain relievers. These items serve various purposes, such as stopping bleeding, preventing infection, and alleviating discomfort. Bandages come in many sizes and styles, from adhesive strips for minor cuts to larger gauze pads for more serious wounds. Diversity is always valuable. Antiseptics help clean injuries and reduce the risk of infection, while pain relievers provide comfort during recovery.
- **Make it adaptable:** A good first aid kit should be adaptable to suit specific family needs. If you have children, you might need pediatric-specific supplies, like smaller bandages or infant medications. It's also crucial to include personal medications prescribed for ongoing health management. Customizing your kit ensures you have what you need in an emergency, reducing stress and enhancing readiness.
- **Include prescription medication:** Prescription medications are critical components of a holistic first aid kit, especially for those managing chronic conditions. Keeping these medications in your kit ensures they are readily available when you may not have access to your usual belongings. Equally important is keeping an updated list of personal medical information. This includes details on allergies, any chronic conditions, and emergency contact numbers. Having this information on hand can expedite care and ensure appropriate treatments are administered quickly.
- **Store your first aid kit optimally:** Storing your first aid kit in an accessible location with clear labeling is vital for ease of use during emergencies. Whether in the home, car, or workplace, knowing exactly where the kit is located can save valuable time. Clear labeling helps identify the contents swiftly so family members or caregivers can find the right supplies without delay. It's also important to ensure that everyone in your household understands where the kit is and how to use it, enhancing overall preparedness.
- **Regularly check and update your kit:** Regularly checking your first aid kit is another key aspect of maintaining its effectiveness. Supplies can expire or become depleted over time, so periodic reviews are necessary to ensure everything is in good condition. This involves examining items for expiration dates, replenishing used materials, and updating personalized content according to current health needs.

This maintenance keeps the kit ready for emergencies and also provides peace of mind, knowing you are prepared for unforeseen situations.

- **Consider environmental factors:** It is also beneficial to consider environmental factors when assembling your first aid kit. For instance, if you live in a region prone to extreme weather conditions, you may want to include specific items like a thermal blanket or sun protection. Adjusting your kit to geographic and lifestyle factors further enhances its utility, providing a tailored solution for the unique challenges your environment may pose.

Once you've put together the perfect first aid kit, you need to educate your whole family so they know exactly how to use it in the event of an emergency. Simple training sessions can ensure everyone knows how to apply bandages, use antiseptics, and administer medications. This education is particularly important for older children and teenagers who might find themselves in situations where they need to assist in an emergency.

Training in First Aid Techniques

Understanding the importance of first aid training is essential for anyone who values safety and preparedness in emergency situations. Training in cardiopulmonary resuscitation (CPR) and basic first aid equips you with the skills necessary to respond swiftly and effectively to medical emergencies. The following points showcase how proper training guarantees the effective use of first aid supplies while encouraging confidence and invaluable knowledge:

- **Participate in CPR and first aid courses:** Everyone should enroll in CPR and first aid courses to empower them during emergencies. These courses cover a range of emergency scenarios, offering you and your family insights into techniques that can make a significant difference when seconds count. For example, learning CPR involves understanding how to perform chest compressions to maintain blood circulation when someone is experiencing cardiac arrest. Immediate CPR can double or even triple the chances of survival following such an event. Beyond heart attacks, these courses cover common incidents like choking and drowning, where prompt intervention can prevent catastrophic outcomes.
- **Practice real-life scenarios:** Practice scenarios offered in these classes help you build confidence and enhance skill retention. You will most likely be placed in simulated emergency situations where you must apply what you've learned. These practical exercises, conducted under controlled conditions, allow you to practice decision-making, quick thinking, and hands-on application of first aid procedures. When faced with real-life emergencies, this experience inspires calmness and decisive action. Through repetition and reinforcement, trainees are better prepared to recall necessary procedures and equipment usage without hesitation.

- **Take quick action:** Quick action during emergencies is crucial. The window of opportunity to provide life-saving interventions can be as short as a few minutes. CPR classes teach you how to act rapidly and accurately. They highlight the importance of assessing the situation quickly, calling for professional medical assistance, and beginning CPR immediately if needed. Fast response increases the chances of survival while reducing the severity of injuries by maintaining oxygen flow to vital organs until professional help arrives.
- **Cover various possible scenarios:** Training covers a wide array of possible incidents, enhancing your general preparedness beyond specific situations. You learn to handle various medical emergencies, such as severe bleeding, burns, fractures, and allergic reactions. Understanding how to administer first aid for these issues prepares you to address different needs using available resources. Through this training, you can also learn how to use automated external defibrillators (AEDs), which can help you address sudden cardiac arrests, exponentially increasing the likelihood of positive outcomes.
- **Understand each procedure in depth:** The value of first aid training extends beyond just knowing the steps. It includes understanding the rationale behind each procedure, which aids in adapting responses to unique circumstances. With comprehensive training, you can be prepared for both personal emergencies and assisting family members, neighbors, or even strangers in public settings. This preparation encourages a community-oriented mindset, where more people are equipped to assist each other when in need.

Practicing this training can help you be prepared for any medical emergency that may come your way. It's valuable for everyone in the home to become prepared for various types of events.

## Management of Over-The-Counter and Prescription Medications

If you take prescription medications, knowing how to manage them effectively during an emergency is vital. It's also valuable to prepare yourself with all the over-the-counter medicine you may use regularly. When discussing the management of essential over-the-counter (OTC) and prescription medications in households, it is crucial to recognize their role in ensuring both immediate relief and long-term health management. When preparing for medical emergencies, it's necessary to understand which medications are essential. The following tips can help you manage your different types of medications:

- **Stock up on essential OTC medicines:** Essential OTC medications form the first line of defense against common ailments. Pain relievers like ibuprofen or acetaminophen are staples, providing quick relief from headaches, muscle aches, and minor injuries. Antihistamines such as diphenhydramine or loratadine address allergies, offering relief from itchy eyes, runny noses, or hives. Digestive aids, including antacids or antidiarrheals, help you manage upset stomachs or diarrhea, which often require prompt attention. By keeping a well-stocked supply of these medications, you can handle minor health issues swiftly, reducing the need for unnecessary medical visits. You can find a list of the top 10 essential medications you should stock at the end of this section.

- **Stockpile prescription medication:** For prescription medications, responsible stockpiling can help you avoid interruption in treatment. Consider communicating with health care providers about obtaining an extra supply, particularly for life-dependent medications like blood thinners, insulin, or asthma inhalers. A prudent approach is maintaining a buffer supply—enough to cover unforeseen circumstances such as supply chain disruptions or natural disasters. This practice alleviates anxiety about running out during critical times.

- **Prepare for any chronic conditions:** If you or any of your family members has a chronic condition such as diabetes or asthma, use strategies that focus on consistent medication adherence and proactive health monitoring. For diabetics, this means having adequate supplies of insulin, syringes, test strips, and glucose tablets. It's equally important to educate family members on recognizing symptoms of hypo or hyperglycemia and knowing how to administer emergency treatments if necessary. Asthma sufferers should maintain accessibility to rescue inhalers and nebulizers, especially during high pollen seasons or periods of increased air pollution. Regular check-ins with health care professionals can optimize treatment plans and adjust medications according to evolving health needs.

- **Keep a list of current medications:** Maintaining a current record of medications, dosages, and treatment schedules can be lifesaving during emergencies. An updated list stored in a readily accessible location enables caregivers or first responders to quickly understand a patient's medical history, preventing adverse drug interactions and ensuring continuity of care. It's advisable to include details of the prescribing physician and pharmacy contact information for further reference during emergency situations.

- **Inform family members of the medicine stockpile:** Educating all household members about the proper storage and handling of medications is another key aspect of effective management. Medications should be kept in a cool, dry place, away from direct sunlight, to preserve their efficacy. Furthermore, establishing routines, such as setting alarms or organizing pillboxes, can assist in maintaining regularity in taking medications. Involving technology, like smartphone apps, can provide timely reminders and track compliance.

It's important to develop an environment of open communication within your family regarding health matters to enhance overall preparedness. Discussing each member's medical needs and any changes in their condition encourages mutual support and shared responsibility, building a reliable framework for managing health collectively.

## *10 Essential Medications for Emergencies*

When preparing for emergencies, it is crucial to have a well-stocked medicine cabinet. Having essential medications on hand can make a significant difference in managing situations, whether dealing with minor injuries or more severe health concerns. Here are 10 medications that everyone should consider keeping readily available.

### Pain Relievers

Over-the-counter pain relievers, such as ibuprofen or acetaminophen, are important for treating minor aches and pains. Ibuprofen can also help reduce inflammation and fever. For example, if someone has a headache or muscle pain after an accident, taking ibuprofen can provide quick relief. Always check the dosage on the label, especially for children, to ensure safety.

### Antihistamines

Antihistamines like diphenhydramine are essential for managing allergies and allergic reactions. They can be useful during spring and summer when pollen counts are high. Besides, if someone gets stung by a bee or has a mild allergic reaction to a food item, antihistamines can alleviate symptoms like itching, hives, or swelling. It is beneficial to keep both the regular and nondrowsy formulas available.

### Cold and Cough Medications

Cold and cough medications can help relieve symptoms associated with respiratory infections. Look for expectorants to help loosen mucus or cough suppressants for dry coughs. For instance, if a family member catches a cold, taking cough syrup can help them sleep better by calming their cough during the night. Read the instructions carefully and consider consulting a health care provider if symptoms persist.

### Antacids

Antacids, such as calcium carbonate or magnesium hydroxide, are useful for sudden digestive issues like heartburn or indigestion. They work by neutralizing stomach acid. If someone eats too much rich food or has spicy meals, antacids can quickly soothe the discomfort. Keeping chewable tablets on hand is a practical option for instant relief.

### Antibiotic Ointment

An antibiotic ointment like bacitracin or Neosporin is crucial for treating minor cuts and scrapes. These ointments prevent infection by keeping the wound moisturized and creating a barrier. For example, after a small fall, cleaning the wound and applying antibiotic ointment can help it heal properly without complications. It is essential to keep a tube in your first aid kit.

**Hydrocortisone Cream**

Hydrocortisone cream is beneficial for reducing inflammation and treating skin irritations. It can relieve symptoms from bug bites, rashes, or eczema flare-ups. If someone gets bitten by mosquitoes during a picnic, applying hydrocortisone cream can significantly reduce itching and swelling. Having this cream available can provide a quick solution for various skin issues.

**Electrolyte Solutions**

Electrolyte solutions, like those available in powder form, can be very helpful during dehydration, especially after severe vomiting or diarrhea. These solutions replenish lost fluids and electrolytes. For instance, if someone has a stomach virus, drinking an electrolyte solution can help restore their body's balance and speed up recovery. It is wise to keep packets in your emergency supply kit.

**Growth Hormone Injections**

For those individuals with diabetes, having insulin or other prescribed growth hormone injections on hand is critical. In emergencies, insulin can help lower blood sugar levels if they rise too much. Always ensure you understand how to administer these injections correctly and keep them in a temperature-controlled environment to maintain their effectiveness.

**Family Prescription Medications**

Stocking essential prescription medications is necessary for those with chronic health conditions. For example, if a family member relies on medications for conditions like high blood pressure or asthma, having an extra supply ensures you are prepared for emergencies. Consult with a health care provider about the best practices for storing and managing these medications.

**First Aid Kit**

Lastly, while not a medication, having a comprehensive first aid kit is essential. Your kit should include Band-Aids, gauze, medical tape, scissors, tweezers, and additional items. In any emergency, having a first aid kit allows you to treat injuries on the spot. For example, if someone cuts themself while cooking, you can quickly clean the wound and apply a bandage.

Being prepared for emergencies involves having the right medications and supplies readily available. Stocking your cabinet with these ten essential medications and items will help you handle unexpected situations better.

Make it a habit to periodically check expiration dates and restock as necessary. This practice ensures that when emergencies arise, you will have the tools needed to manage them effectively.

Natural Remedies and Self-Care for Common Ailments

You may not want to rely on medical supplies, or you might want healthy alternatives that can treat various medical conditions during unplanned events. In the face of medical emergencies, natural remedies can serve as valuable backup solutions for common health issues. While these remedies should not replace professional medical treatment, they offer complementary benefits that can be integrated into a holistic approach to health care. Here are various reasons why herbal and natural remedies are a great option to explore:

- **Treating cold symptoms:** Herbal treatments serve as a first line of defense against minor ailments, such as colds. When it comes to easing cold symptoms, certain herbs can be effective. Echinacea, for example, is believed to boost the immune system and might reduce the duration of cold symptoms by 10%–30% (Editorial Contributors, n.d.). Elderberry is another herb known for its potential antiviral properties, which could help alleviate flu-like symptoms more quickly (Editorial Contributors, n.d.). Herbal teas, including chamomile and peppermint, are also popular choices, offering soothing effects that can relieve congestion and sore throats.
- **Treating headaches:** Headaches are another common issue that can be successfully managed using herbal remedies. Peppermint oil, when applied topically, has a cooling effect that may relieve tension-type headaches. Ginger tea is also helpful as it can reduce inflammation, providing comfort for headache sufferers. Moreover, acupuncture, an ancient Chinese practice involving the insertion of thin needles into specific points of the body, has shown promise in reducing headache frequency and severity (Kubala, 2018). These traditional methods highlight the potential of natural remedies to ease discomfort without the side effects often associated with conventional medications.
- **Combination with traditional medicine:** Complementing traditional medicine with holistic approaches offers a broader perspective on health management. To do this successfully, you should consider yourself as a whole—mind, body, and spirit—rather than focusing solely on symptoms. Integrative medicine practitioners often recommend using natural therapies alongside pharmaceuticals to enhance overall wellness. Incorporating holistic strategies like yoga and meditation can promote relaxation and improve mental health, which in turn supports physical healing. Yoga, for instance, not only increases flexibility but

also helps reduce stress levels, ultimately leading to a decreased frequency of headaches (Kubala, 2018).

- **Managing stress:** Stress, a significant contributor to numerous health problems, can be managed through basic techniques like breathing exercises and mindfulness meditation. Deep breathing promotes relaxation by activating the parasympathetic nervous system, helping to decrease stress hormones in the body. Mindfulness meditation encourages individuals to focus on the present moment, fostering a sense of calm and reducing anxiety. Regular practice of these techniques can significantly enhance mental health, providing individuals with tools to navigate life's stresses more effectively.
- **Staying safe:** When using natural remedies, it is essential to practice safety and sensibility. Combining natural treatments with mainstream medications requires careful consideration of potential interactions. For instance, certain herbs can affect the metabolism of drugs, either enhancing or diminishing their effectiveness. It is crucial to consult health care professionals before introducing new remedies, especially if other medications are being taken. This ensures that natural and pharmaceutical treatments work harmoniously without compromising health.
- **Being mindful of allergies:** Maintaining awareness of possible allergies to herbal ingredients is a key aspect of safe practice. Allergic reactions can occasionally occur, emphasizing the importance of conducting patch tests or consulting with health care providers when trying a new remedy. Moreover, natural products should always be sourced from reputable suppliers to ensure quality and efficacy, as the market is saturated with varying standards of preparation.

If you're a parent or caregiver, you should stay informed about the appropriate use of these treatments to safeguard family health during minor medical situations. Understanding dosage, possible side effects, and the correct application of natural remedies is critical in maximizing their benefits while minimizing risks.

Special Considerations for Families and Unique Needs

It's important to remember that every family is different, so your family may have specific medical needs that you should prepare for. Taking each family member's medical needs, age, and health conditions into consideration can help you take into account everyone's personal medical requirements. Here are some different factors to consider when achieving medical preparedness:

- **Allergies:** Allergens pose serious risks, especially for those with severe reactions. Identifying potential allergens in your environment is the first step toward prevention. This means knowing common triggers such as foods, insect stings, or medications that could cause anaphylaxis. Once identified, it's imperative to have measures in place. For instance, carrying antihistamines or epinephrine auto-injectors can form an integral part of this preparedness. In addition, it's beneficial to inform family members or close contacts about these allergens and educate them on recognizing symptoms and administering life-saving medication quickly. This collaborative awareness creates a safety net around the affected individual.

- **Chronic health conditions:** For individuals with chronic health conditions, planning revolves around ensuring continuity of care even when regular health care access might be disrupted. Conditions like diabetes, hypertension, or asthma require consistent management; thus, having a well-stocked supply of medications and necessary equipment is essential. Diabetics should ensure they have enough insulin and blood glucose monitoring supplies, while asthmatics need inhalers within easy reach. It's also advisable to routinely check expiration dates and replace meds and tools to guarantee their effectiveness during an emergency.

With these factors in consideration, you should follow these tips to use your unique knowledge to better your medical preparedness:

- **Medical information:** Creating a detailed medical information sheet serves as an invaluable resource during crises. This sheet should include critical information like personal identification details, emergency contacts, primary physicians, and a list of current medications alongside their dosages and any allergies. It should also outline particular medical conditions and essential equipment required, such as hearing aids or pacemakers. Keeping this sheet easily accessible—perhaps stored in a waterproof, visible location—ensures that responders or bystanders can provide appropriate assistance promptly if needed. This proactive documentation can be greatly beneficial for medical teams, allowing them to act swiftly.
- **Consider the unique challenges of children or the elderly:** Families with children or elderly members may face additional challenges due to varying physical and cognitive capabilities. As a parent, you should pack comfort items like plush toys or favorite snacks to alleviate stress for younger children who might struggle to process the situation. Meanwhile, consideration for mobility aids or communication devices for the elderly can significantly reduce distress, providing assurance that their needs will not be neglected. Making these adjustments is important for all family members.
- **Promote inclusivity:** Inclusivity is especially important for you and your family. You want to ensure that you're catering to everyone's health and safety, even during emergencies. Ensuring materials cater to dietary restrictions, sensory disabilities, or language preferences—perhaps using preprepared flashcards for nonverbal communication—can bridge gaps that hinder effective response efforts. Developing

these tailored solutions necessitates recognizing each individual's inherent worth and dignity, which integrates empathy within practical strategies for medical readiness.

Once you've got a comprehensive first aid kit and a thorough emergency plan, it's important to make changes and adapt to your family's personal needs.

Summary and Reflections

In this chapter, we've explored the crucial aspects of being medically prepared to manage emergencies effectively. It starts with the significance of having a holistic first aid kit filled with essential items like bandages, antiseptics, and personal medications tailored to your needs. We also discussed the importance of storing these kits in accessible locations and regularly checking them for expired supplies. Alongside physical preparedness, understanding the environmental factors and adapting your first aid kit accordingly enhances readiness. We've also touched on educating all family members about the contents and usage of your kit, ensuring everyone knows how to handle minor medical incidents independently.

The value of first aid training was also emphasized in building confidence and skill retention, highlighting its role in effective emergency response. Engaging in CPR and first aid courses empowers you with knowledge that could save lives during critical moments. We further discussed responsible management of over-the-counter and prescription medications, focusing on stockpiling and adherence to chronic conditions. Finally, we looked at the benefits of integrating natural remedies alongside conventional treatments for common ailments while addressing special considerations for unique family needs. This holistic approach to medical preparedness ensures families feel secure and well-equipped to face unexpected situations confidently. With your medical equipment secured, let's move on to some other essential tools and protective gear you may need to protect yourself and your family in the next chapter.

# 6

# ESSENTIAL SURVIVAL GEAR AND PROTECTIVE EQUIPMENT

Having the right survival gear for any emergency can prepare you for unexpected crises. The unpredictable nature of emergencies, ranging from natural disasters to man-made threats, requires a well-rounded approach to readiness. Throughout this chapter, we'll discuss the various components that form the backbone of any effective survival plan, highlighting the critical importance of selecting and investing wisely in essential tools. Understanding the role of each piece of equipment can make the difference between vulnerability and safety when faced with challenging situations.

Furthermore, you will gain insight into essential survival gear, beginning with basic items such as flashlights, ropes, and compasses, which are integral to navigating and securing safety during emergencies. We'll also discuss the significance of personal protective equipment (PPE), emphasizing the necessity of masks, gloves, and goggles in safeguarding against environmental hazards. The chapter also offers practical advice on tailoring survival kits to meet specific risks depending on geographic and situational factors. For those seeking additional expertise, insights from military strategies provide valuable guidance on the strategic use of tactical gear, including firearms and home security systems.

## Basic Survival Gear Essentials

Having survival gear is essential to feel well-prepared for any crisis or emergency. This focus on preparation isn't just about gathering equipment; it's about understanding and investing in what truly matters for different scenarios. These are some of the most foundational items you should keep to ensure your security and well-being:

- *Flashlights* are simple yet vital tools that provide you with illumination when natural light and electricity are not an option. Whether navigating through the dark corridors of your home during a blackout or trekking across uncertain terrain at night, a reliable flashlight ensures visibility and security. In this setting, consider opting for LED models. These offer longevity and efficiency, while solar or hand-crank options guarantee you're not left without light when batteries deplete. A small flashlight can guide you through the darkness, prevent fall hazards, and even signal for help when necessary.

- *Ropes* provide versatile functionality in your toolkit. They are indispensable for tasks such as securing shelter, creating makeshift rescue lines, or even descending steep inclines. A strong, durable rope can bridge the gap between safety and vulnerability, especially when nature transforms familiar surroundings into treacherous challenges. Investing in high-quality climbing ropes can ensure they withstand significant tension and abrasion, providing peace of mind when every second counts.
- *Compasses* serve as crucial navigational tools, offering reliable guidance when technology fails. Unlike GPS devices, which depend on satellite signals and power sources, a compass remains unfailingly reliable. During emergencies, when modern conveniences are unavailable, knowing how to read a compass can mean the difference between finding your way to safety and becoming disoriented. Paired with detailed maps, compasses can guide you through wild terrains or chart routes around obstacles despite any challenging conditions you may be in.

With these basic necessities in your emergency kit, it's important to practice the following tips to be ready for anything:

- **Consider space-saving:** Efficiency and space-saving are important considerations when putting together your crisis kit. Here, multipurpose gear stands out for its ability to amalgamate multiple functions within single tools, minimizing bulk and maximizing utility. Items such as Swiss Army knives, which come equipped with blades, screwdrivers, scissors, and more, exemplify this concept. Similarly, compact multitools combine several utensils into one device, eliminating the need for carrying multiple separate items. By choosing multipurpose gear, you can keep your kits manageable while ensuring no functionality is sacrificed.
- **Pack location-specific gear:** Sometimes, basic gear doesn't meet all your survival needs. Location-specific gear, tailored to anticipated environmental and situational demands, enhances preparation significantly. For individuals living in flood-prone areas, inflatable life vests or waterproof bags are crucial additions to their usual gear. Those in earthquake zones might prioritize sturdy shoes and gloves to navigate debris safely. Thus, understanding local risks allows for better customization of survival kits, ensuring they address specific threats effectively.

- **Organize your gear:** Organizing and prioritizing gear within kits helps you navigate it in stressful and fast-paced situations. Having a neat and accessible arrangement can improve your readiness when faced with urgent situations. Strategically placing critical items toward the top or in external pockets can save valuable time during evacuation or emergency response. Prioritizing light, communication tools, and first aid essentials ensures these vital components are within quick reach. In addition, developing an inventory checklist helps maintain organization over time, bringing your attention to any missing or expired items that require replacement.

While it's imperative to assemble a kit, it's equally important to enrich yourself with the knowledge and practice to use this gear effectively. You should familiarize yourself with operational procedures, from reading maps and compasses to deploying shelter and utilizing ropes safely.

Personal Protective Equipment (PPE)

To prepare yourself for different situations, it's important to keep protective equipment that can ultimately be life-saving. As you keep prioritizing safety and preparedness, it's crucial to understand how to procure and utilize PPE effectively.

Here's a guide you can follow when gathering and using PPE:

- **Types of PPE to purchase:** Masks, gloves, and goggles form the frontline defense against various environmental hazards. Masks, such as N95 respirators or surgical ones, are essential during events like pandemics or smoke from fires, filtering out harmful particles and pathogens. Always ensure that masks fit snugly over your nose and mouth without gaps to maximize effectiveness. Gloves protect against chemical spills and biological contaminants, preventing skin contact with dangerous substances. When selecting gloves, consider the material: Nitrile gloves are resistant to many chemicals, whereas vinyl gloves might suit other tasks better. Goggles shield the eyes from debris and chemical splashes. Look for goggles meeting ANSI standards, making sure they have no cracks or deformities, and that the straps securely seal to the face (Tarlengco, 2024).
- **Invest in good quality:** Sturdy and durable PPE is important, even though it might cost a bit more. In emergencies, such as natural disasters or industrial accidents, the right attire can minimize risks of injury. Sturdy clothing, like fire-resistant suits or waterproof jackets, offers protection from extreme temperatures and weather conditions. Consider gear made from durable materials like Kevlar or Gore-Tex for maximum resilience. Footwear, including steel-toed boots and anti-slip shoes, is also vital for navigating through debris or slippery surfaces safely. They

provide essential support and protection, reducing the risk of foot injuries, which could hinder movement during critical moments.

- **Store your PPE accessibly:** The accessibility of protective gear is a key factor because of its quick deployment when needed most. It is advisable to store PPE in an easily reachable location within your home or workplace. For instance, a dedicated emergency kit near exits or common areas can facilitate rapid access. Regular drills can help familiarize everyone with the locations and proper usage of PPE, minimizing panic during actual emergencies. Having spare items, especially commonly used ones like masks and gloves, guarantees that you won't run short in prolonged crises.
- **Get a proper fit:** Make sure your PPE fits properly before purchasing it. Ill-fitting PPE can leave gaps in protection, exposing you to potential hazards. Conducting regular fit checks, especially for masks and gloves, ensures they keep their protective seal. It's recommended that every member of the household tries on each piece of equipment before an emergency arises. Moreover, maintenance extends the lifespan and efficacy of your protective gear. Clean and disinfect items like goggles after each use, following manufacturer instructions to prevent damage (*Personal Protective Equipment*, 2023). Be mindful of how you store your PPE. Store them away from direct sunlight or harsh environments that can degrade materials over time.

Take good care of your PPE, and they can last forever. Although this equipment is usually used very seldomly and sometimes never touched at all, it's good to have in the unlikely event of a crisis. Remember to always follow the manufacturer's instructions when using your PPE.

Customized Equipment Based on Risk

If you are focused on more high-risk emergencies, you can't go wrong with customizing your equipment based on risk. Knowing what hazards you might face and understanding how best to prepare for them can make all the difference in an emergency situation. Here are some tips to consider when customizing your own equipment:

- **Practice situational awareness:** Situational awareness plays a critical role in shaping your survival equipment choices. Whether you live in a region prone to natural disasters like hurricanes or earthquakes or simply want to be prepared for any crisis, being aware of potential threats helps determine the right gear. For example, if you reside in an area susceptible to wildfires, investing in fire-retardant clothing, respirators, and a reliable water source becomes a priority. On the other hand, residents of flood-prone areas should look into waterproof gear, such as dry bags and life vests, to ensure safety during unexpected water surges. These more

specific preparations help you prepare for disasters that are more likely to occur in your environment.
- **Cater to specific threats:** Catering gear selection to specific threats adds another layer of readiness. For instance, regions that experience harsh winters require different survival tools compared to those with warmer climates. Snow shovels, thermal blankets, and emergency heating sources are crucial in cold environments, while heat resistant materials and hydration packs serve well in hot climates. In more urban settings, equipping yourself against man-made threats might involve different considerations, such as air filtration masks to protect against pollutants or collapsible ladders for safe evacuation from multistory buildings. Thinking through these specific scenarios and customizing your gear accordingly can significantly enhance your ability to respond effectively to emergencies.
- **Consider your pets:** Including provisions for pets is equally important. Pets are members of the family and should have their own supplies ready, including food, water, medications, and comfort items like blankets and toys.

With a well-thought-out plan, it's important to continue maintaining and adapting it to meet your changing needs.

Maintaining and Checking Gear Readiness

Ensuring that your survival gear is always ready for use is a critical element of effective crisis management. Practice the following steps to ensure your gear is always ready for anything:

1. **Conduct regular inspections:** The first step in achieving this level of preparedness involves regular inspections and timely replacements of any damaged or outdated items. Just like any equipment used frequently, survival gear can wear out over time due to environmental factors or simple usage during training sessions. Instead of waiting until something breaks at an inconvenient moment, systematically check each piece of gear every six months. This includes inspecting your gear regularly for physical damage such as rust, tears, or cracks and also ensuring that devices like flashlights and radios are operational. Consistently replacing worn or damaged items helps prevent minor issues from escalating into significant problems, guaranteeing the reliability of your equipment when it's most needed.
2. **Educate yourself and your family:** To utilize your survival gear effectively, you should educate yourself and your family. Training programs designed to teach proper usage and techniques can provide you with invaluable skills. For instance, knowing how to properly set up a tent, operate a portable water filter, or handle emergency medical kits can make all the difference in a crisis. Many organizations offer workshops and courses that cover the fundamentals of using different types of

survival equipment. Even watching online tutorials can be beneficial; however, nothing beats hands-on experience under professional supervision.

3. **Create an inventory checklist:** To keep your gear organized and accessible, developing an inventory checklist is highly recommended. Start by listing all your essential items, categorizing them by function—such as shelter, food, water, first aid, and tools. This checklist serves as a quick reference guide, allowing you to easily track what you have on hand and identify what's missing or needs replacement. Having a well-organized checklist simplifies the process of checking your supplies, especially when you're stressed or in a hurry. Revisit and update your inventory regularly, ideally every six months, in conjunction with your gear checks, to ensure nothing has expired or gone missing. Digital applications available today can help maintain such lists seamlessly, offering reminders for inspection dates and renewals.

When you maintain your gear well, you'll be able to keep this survival equipment for long. This helps you reduce any wear and tear that requires gear replacements.

Essential Survival Gear Recommended by Elite Military Experts

Elite military operatives have long been regarded as experts in survival and tactical operations, offering invaluable insights into essential gear that can mean the difference between life and death in crisis situations. These are some of their top recommendations that you should consider for your home security with the correct safety precautions:

- **Firearms:** Firearms are crucial for home defense, offering reliable protection against intruders or threats. The choice of your firearm depends on personal preferences, home layout, and your comfort level with weapons. Commonly recommended firearms include shotguns, handguns, and rifles. Shotguns, particularly 12-gauge models, are favored due to their wide pellet spread, especially when used in confined spaces. They are highly effective but require training for effective use. Handguns offer greater maneuverability; models such as Glock, Smith & Wesson, and Sig Sauer provide comfortable handling, reliability, and noteworthy stopping power. Selecting the right ammunition is equally important, with hollow-point bullets often preferred for their reduced risk of over-penetration and increased incapacitation potential.
- **Nonlethal defense tools:** If you're uncomfortable with lethal options, there are a variety of nonlethal defense tools you can look into, such as Tasers and pepper spray. Tasers deliver an electric shock capable of incapacitating an attacker temporarily, allowing you to escape or call for help. Advancements in Taser technology have led to compact, easy-to-use designs suitable for close-range defense. Pepper spray offers another layer of protection, dispersing a powerful irritant that causes temporary

blindness, difficulty breathing, and intense discomfort, effectively deterring or neutralizing would-be attackers. Both Tasers and pepper spray provide you with a sense of security without feeling uncomfortable with lethal weapons.

- **Technological advancements:** Beyond physical defense mechanisms, technological advancements have significantly improved home security systems. A reliable security system typically includes door and window sensors, motion detectors, surveillance cameras, and alarms. These systems can alert you when unauthorized activities happen. Modern systems also provide remote monitoring capabilities through smartphones or tablets, enabling real-time updates regardless of your location. Integration with smart home technology allows for greater control and customization, catering to specific needs or preferences.
- **Tactical gear:** Tactical gear, including body armor, holds paramount importance in high-risk scenarios. This equipment offers protection against sharp objects and blunt impacts. Body armor ranges from lightweight vests to heavier, reinforced plates designed to withstand rifle rounds. It is crucial to select the appropriate level based on perceived threats and personal mobility needs. While not everyone may anticipate needing body armor, having access could drastically enhance your safety in emergencies. In addition to body armor, other tactical gear includes utility belts, holsters, and backpacks designed to carry essential items efficiently. These items facilitate organization, ensuring quick access to necessary equipment when seconds matter. For example, duty belts with custom compartments allow users to store firearms, magazines, communication devices, and other critical tools systematically.

When considering survival gear recommended by military experts, it is essential to recognize that investment in quality equipment directly correlates with enhanced safety and preparedness. You should assess your unique circumstances and requirements to determine what combination of lethal and nonlethal tools effectively suits your needs. Note that regular training ensures proficiency with chosen defensive measures, allowing you to use your equipment safely and confidently.

Final Thoughts

In this chapter, we've walked through the essential survival gear and protective equipment needed to face various crises effectively. Key items such as flashlights, ropes, and compasses are invaluable when natural conditions become unpredictable, providing light, support, and reliable navigation. Our focus has also covered multipurpose gear that combines efficiency and space-saving capabilities, ensuring preparedness without unnecessary bulk. Customizing your gear based on specific risks enhances its effectiveness, allowing you to adapt to local threats while taking into account personal and family needs. This unique approach guarantees readiness whether you're facing a natural disaster or a man-made crisis.

The importance of personal protective equipment (PPE) was also discussed, with emphasis on how masks, gloves, goggles, and sturdy clothing play crucial roles in safeguarding against environmental hazards. By understanding the necessity of each item and practicing its use, you build the confidence to act decisively in emergencies. As we conclude this discussion, it's clear that investing in the right gear is not just about gathering items; it's about being prepared to protect yourself and your loved ones in any situation. With the right survival gear, you can always be a step ahead when a crisis happens. The next step for preparation includes planning ahead for the financial and legal impacts of crisis and emergency management. This will be covered in depth throughout the next chapter.

# 7

# FINANCIAL AND LEGAL PREPAREDNESS FOR CRISIS SITUATIONS

In the event of an emergency or crisis, you may overlook the financial and legal impact, but it's important to consider these implications. Financial and legal preparedness can help you handle any situation more holistically. While much attention tends to focus on immediate physical safety and resource management, the financial and legal aspects can enhance your stability and resilience during challenging times. When disaster strikes, be it natural, economic, or health-related, having a solid foundation in these areas can make a substantial difference in your ability to effectively navigate the uncertainties that accompany different crises. By understanding and addressing these often-complex considerations, you can survive and recover from emergencies with more ease.

Within this chapter, you will learn about various financial and legal strategies crucial for facing crises head-on. The discussion explores practical approaches to managing finances, including establishing budgets focused on essential needs and leveraging resources like insurance and diversified investments for asset protection. You'll understand the significance of legal knowledge, touching on topics such as tenant rights and contractual clauses that may come into play during emergencies. Homeowners, renters, and families should all have access to this knowledge so they can develop informed plans that incorporate both financial security and legal awareness.

## Managing Finances During a Crisis

When you're in a crisis, the last thing you may be thinking of is your finances. Often, during emergencies, your finances can become essential for ensuring that you and your family remain secure and well-provided. Crises such as natural disasters, economic downturns, or health pandemics can disrupt the usual flow of life, meaning you have to be prepared with viable strategies to manage your financial resources wisely. These tips can help you strategically manage your finances:

- **Create a budget:** Developing a budget that prioritizes essential supplies is the first step in ensuring resource availability when a crisis hits. Consider all the essential supplies discussed in previous chapters, such as food, water, medical supplies, and basic utilities. Creating a detailed list of these items can help you focus financial efforts in the most necessary areas. One tip is to categorize expenses into needs versus wants. Essential supplies should always align with genuine needs, which ensures money is spent wisely. Make use of budgeting tools or apps to track and adjust spending as your circumstances evolve. Having these plans can help you stay ahead of your finances.
- **Stretch your supplies:** Another important strategy is to implement resource rationing to stretch existing supplies over an uncertain period. You should prolong the usability of available resources so that you have enough to last you through unforeseen circumstances. Follow these steps to prepare and stretch your supplies successfully:
  - Begin by assessing current inventory levels within your household, including pantry items, medication, and hygiene products.
  - Calculate how long these supplies might last under normal usage and devise a consumption plan that extends their viability. For instance, adopting measures such as preparing meals in bulk, reducing the frequency of use of certain nonessential items, and limiting waste can conserve resources effectively.
  - Aside from tangible goods, consider rationing energy and water, which can not only cut costs but also ensure continued access through lean periods.
- **Prioritize financial security:** Having financial security during emergencies can reduce your stress levels. During stable times, continually contributing to an emergency fund can cushion the blow of an unexpected crisis. Ideally, this fund should cover at least three to six months' worth of living expenses. If building such a reserve seems daunting, start small and increase contributions gradually. Assess your monthly income and identify areas where spending can be reduced to boost savings. Automating these savings directly from your paycheck into a dedicated account can also strengthen your safety net. Note that emergency funds serve dual purposes; they protect against immediate unforeseen expenses and reduce stress associated with financial instability.
- **Diversify your accounts:** Diversification in financial planning is valuable for protection during crises. While traditional savings accounts are fundamental, consider alternative saving avenues such as low-risk investments that offer liquidity.

Diversified portfolios withstand market volatility better than singular approaches and may offer additional returns to bolster your emergency reserves.

While every crisis comes with its distinct challenges, having a detailed financial plan can mitigate much of the unpredictability involved. Don't forget to include your family members in these financial plans and decisions so that they're also covered during emergencies. Educating your loved ones on the importance of gathering financial resources for a rainy day or emergency can be helpful to them.

Protecting Your Assets

It's always important to safeguard your financial assets, especially when you're faced with crisis situations. It's not just about having a plan but ensuring that these crucial elements are protected in every possible way. This involves securing your home and possessions, diversifying investments to weather economic storms, and maintaining access to financial resources when they are needed most. The following steps should be taken into consideration when working toward securing your assets:

- **Secure your physical property:** To begin with, securing physical property and valuables should be prioritized through diligent documentation and comprehensive insurance coverage. Start by creating a detailed inventory of all valuable items in your possession, including electronics, jewelry, antiques, and important documents. Photographs or videos can serve as visual proof of ownership and condition, which can be invaluable during insurance claims. Store these records securely, preferably in cloud storage or a fireproof safe.
- **Cover your assets with insurance:** Insurance is another critical component in protecting physical assets. Let's take a closer look into the value of insurance and how it can be helpful. Homeowners and renters should ensure their policies adequately cover unexpected damages from natural disasters, theft, or accidents. Therefore, evaluate your current policy to determine if additional riders are necessary for high-value items. For instance, in regions prone to flooding or earthquakes, it is wise to consider specialized coverage. Insurance provides peace of mind and financial support in rebuilding or replacing damaged assets, ultimately reducing stress during crises.
- **Frequently assess and rebalance your portfolios:** It's essential to regularly assess and rebalance portfolios. This proactive approach ensures investments align with changing personal goals and market conditions. Professional financial advice can be invaluable here, offering insights based on current trends and future projections. Advisors can help craft strategies tailored to individual risk tolerance and long-term objectives. Additionally, keeping abreast of market developments

empowers investors to make informed decisions, further securing their financial footing.

- **Acquire access to necessary financial resources:** Establishing access to essential financial resources like credit or emergency accounts is crucial when traditional income sources might be disrupted. An emergency fund acts as a lifeline during crises, providing liquidity for unexpected expenses. Financial experts commonly recommend setting aside three to six months' worth of living expenses. These funds should be easily accessible, such as in a savings account that earns interest without restrictive withdrawal penalties.
- **Consider legal aspects:** Lastly, it's important to acknowledge legal frameworks surrounding asset protection. One emerging trend is the use of trusts to safeguard family wealth from creditors or legal actions. Trusts, whether revocable or irrevocable, can offer layers of protection while allowing control over asset distribution. This is particularly useful for individuals with substantial estates or those in professions vulnerable to litigation.

Understanding the nuances of asset protection laws enhances your ability to take advantage of available safeguards. Consultations with legal professionals experienced in estate planning or asset protection can showcase strategies best suited to your personal circumstances.

### Cash, Digital Currency, and Bartering

Having access to diverse sources of finances can ensure you are covered in any scenario. In times of crisis, the usual financial systems we rely on can become less accessible or even completely unavailable. Therefore, understanding alternative forms of currency and exchange becomes crucial. These are different ways you can keep your finances:

- **Cash:** During a crisis, having cash on hand is a critical consideration. Cash, unlike digital transactions that require functioning electronic systems, remains usable when technology fails or power outages occur. For example, natural disasters such as hurricanes or earthquakes often disrupt electricity grids and banking infrastructure. In these scenarios, people without access to physical cash may find themselves unable to purchase essential goods as electronic payment systems are disabled. Prepare by setting aside a reserve of cash for emergencies. Furthermore, cash doesn't depend on external systems to function, making it reliably tangible. Its value is universally recognized, and it can't be hacked or lost in a system crash. Keeping a small amount of cash ensures you are prepared for unplanned interruptions to electronic banking services. It's important to consider how much cash is adequate for your needs, taking into account potential necessities like fuel, food, or medication.

- **Digital currency:** In our increasingly digital era, it's essential to look beyond traditional money. Cryptocurrencies offer a compelling alternative when conventional banking becomes inaccessible. Unlike fiat currency, which is controlled by governments and central banks, cryptocurrencies like Bitcoin operate on a decentralized network. Transactions occur directly between parties with no need for an intermediary, such as a bank. This independence from the traditional banking system can prove invaluable during crises when accessibility to banking services is compromised. Educating yourself about digital wallets and secure ways to store cryptocurrency keys is essential for utilizing this alternative effectively. Some communities have adopted their own local digital currencies, creating systems that function parallel to national monetary systems. These local systems can encourage economic resilience by keeping assets circulating within the community and supporting local businesses during broader financial disruptions.
- **Bartering:** Beyond cash and digital currencies, bartering remains one of humanity's oldest methods of exchange and continues to hold relevance during crises. Bartering involves trading goods and services directly without money changing hands. Historically, this method has been relied upon during periods when official currency systems were unstable or unavailable. Communities employ barter systems by reciprocating skills or items based on mutual need and agreement, circumventing the need for traditional currency. For example, during severe financial downturns, neighbors might trade surplus produce for household repairs. Bartering enables people to obtain what they need without depleting limited cash reserves, conserving resources for other payments that require cash. Barter networks or clubs can further streamline this process by organizing exchanges among larger groups, expanding opportunities for obtaining diverse goods and services.

Each form of alternative currency and exchange discussed here offers unique advantages and challenges. It's essential to evaluate personal circumstances to determine the best approach or combination for your personal financial preparedness. Maintaining this flexibility can help you enhance your financial resilience in unpredictable environments.

Insights and Implications

This chapter has explored the often-overlooked financial and legal aspects of crisis management. We've discussed practical strategies for managing finances, such as budgeting for essential supplies, building emergency funds, and diversifying investments to cushion against unexpected events. The importance of preparing for financial uncertainties was emphasized through examples that highlighted the value of maintaining liquidity via cash reserves, digital currencies, and barter systems. These tools ensure continuous access to necessities even when traditional financial systems fail, keeping you secure in even the most stressful crises.

The role of insurance in safeguarding assets and the use of trusts to protect family wealth are critical aspects of a thorough crisis management plan. By gaining insight into these financial and legal elements, you can feel more secure and prepared to handle the multifaceted challenges of crises, ensuring your safety and stability in uncertain times. With all of your physical and tangible preparedness techniques addressed, it's important to focus on the psychological and emotional impact stressful crisis events can have on you. In the following chapter, we will discuss ways for you to mentally prepare for different emergencies.

# 8

# PSYCHOLOGICAL PREPAREDNESS AND RESILIENCE

Psychological preparedness and resilience are important when approaching challenging situations with confidence. Let's discuss how you can prepare yourself mentally and emotionally so that you can handle the stress of an emergency.

Understanding how our psychological state influences behavior is crucial, particularly when faced with uncertainty or danger. By cultivating mental resilience, you can manage your emotional responses better, enhancing your decision-making under pressure.

Through this chapter, we will discuss a variety of strategies designed to enhance resilience and readiness for emergencies. It begins with an analysis of typical stress responses and illustrates how awareness of these can prevent panic. Techniques such as controlled breathing and mindfulness exercises are introduced as practical methods for calming the mind. We'll also cover precrisis preparations, shedding light on how prior planning and rehearsal bolster confidence and reduce anxiety during actual events. Besides, interpersonal dynamics, including communication and trust within family units, are highlighted as key components that strengthen collective resilience.

## Understanding Your Stress Responses During Emergencies

During emergencies, our bodies can react in very different ways. Being able to understand your body's response to stress can help you maintain calm during these overwhelming situations. When confronted with danger or high-pressure situations, the body initiates what is commonly known as the fight-or-flight response. This physiological reaction is driven by signals sent from the amygdala to the hypothalamus, prompting the sympathetic nervous system to release hormones like adrenaline into the bloodstream (LeWine, 2024). Here's how you can identify stress responses:

- **Physical responses:** Manifestations of this response include increased heart rate, rapid breathing, and heightened alertness, all preparatory measures designed to prepare us for immediate action. Recognizing these signs is paramount; without conscious awareness, they can escalate into panic. By acknowledging these physiological changes, you can practice calming techniques such as deep breathing or progressive muscle relaxation to regain control.
- **Emotional responses:** Alongside physical symptoms, emotional responses like fear and anxiety often emerge during emergencies. These emotions are natural but can become troublesome if they spiral into negative thought patterns. It's important to recognize when these feelings begin to cloud your judgment. Regular mental exercises that encourage self-awareness can help manage emotional responses, preventing them from overwhelming your rational mind. For instance, grounding techniques—such as focusing on the surrounding environment—can anchor one back to reality, mitigating spiraling worries (Taylor, 2022).
- **Decision-making skills:** Stress can drastically impact your ability to make decisions. When stress peaks, the ability to think clearly and make sound decisions may diminish. Stress can cause you to rely on impulsive judgments rather than thoughtful reasoning. To counteract this, implementing strategies for rational thinking under pressure is essential. Techniques such as pausing briefly to assess the situation, prioritizing tasks, or recalling similar past experiences can restore clarity and objectivity. For example, breaking down a problem into smaller parts allows for more manageable and focused decision-making.

When you are under this pressure from stress, it's important for you to find ways to manage it that work for you. One efficient way to prepare for crises is through precrisis preparation, which significantly reduces stress levels by minimizing uncertainty. Preparing in advance instills confidence and provides you with knowledge and tools to respond adequately to emergencies. This might include rehearsing emergency plans, familiarizing yourself with safety procedures, or having regular family discussions about what to do in various scenarios. These preparations lessen the shock and confusion of unexpected events, enabling quicker adaptation and protection.

Building and Encouraging Group Trust

Teamwork makes crises less stressful and overwhelming. Building group cohesion through trust activities allows you to lean on each other when challenges arise. Moreover, trust allows group members to feel secure in expressing themselves openly and sharing vulnerabilities without fear of judgment. This kind of genuine communication fosters deeper connections and secures relationships, crucial elements that help in navigating stress as a cohesive unit.

Trust, however, isn't built overnight. It requires consistent efforts and open lines of communication where every individual's voice is valued and heard. When everyone feels heard and understood, they're more likely to contribute meaningfully to problem-solving processes, which enhances overall group efficiency and morale during trying times. A trusted environment encourages people to express their concerns and ideas freely, creating a synergy that encourages quick and effective resolution of issues as they arise. Positive group dynamics also play an essential role in solidifying unity and morale among family or group members. Here are some ways you can create these group dynamics:

- **Create routines of enjoyable activities:** Establishing rituals or activities that everyone enjoys can enhance these dynamics, providing familiarity and comfort during unpredictable circumstances. These shared experiences reinforce the idea that each member's contribution is valued, creating a sense of belonging and teamwork. For example, regularly setting aside time for a family meal or game night can fortify bonds, making it easier to come together and support one another in a crisis.
- **Acknowledge and celebrate each person's contributions:** Creating positive group dynamics often involves acknowledging and celebrating individual contributions. Whether it's recognizing someone's effort in organizing resources or appreciating a member's skill in keeping spirits high, these acknowledgments build confidence and encourage continued engagement. Valued contributions boost morale and also instill a sense of purpose, driving individuals to remain proactive and supportive even under pressure.
- **Practice effective communication:** Effective communication is another critical pillar that prevents misunderstandings and facilitates rapid problem-solving when stress levels are high. Open, honest communication channels allow everyone in the group to express concerns early, reducing the potential for conflicts and fostering a collaborative approach to tackling challenges. It's important that communication be both verbal and nonverbal—reading body language and tone can be just as revealing as words spoken. Doing this ensures that small issues don't turn into larger problems.
- **Have regular check-ins:** To improve communication, it's valuable to have regular check-ins where everyone has the opportunity to share updates, feelings, and any concerns. These forums provide a safe space for airing grievances and receiving assurance, which can be incredibly grounding during turbulent times.

Building this support system can help you handle crises with a sense of calmness and assurance. By focusing on trust, communication, and engagement, groups can cultivate a resilient mindset prepared to face future adversities confidently. As research shows, high-quality relationships and functional support networks are significant predictors of health and well-being, underscoring the importance of investing in these areas proactively (Ozbay et al., 2007).

Techniques for Staying Calm Under Pressure

You may find it manageable to stay calm while preparing for something. Unfortunately, most emergencies happen abruptly, so you can't prepare for them. This is why it's crucial to learn how to stay calm under pressure. Maintaining composure and clarity is vital for making sound decisions. These abilities can be cultivated through a set of practical techniques that empower you to handle stress effectively and enhance your resilience. Here are some of the most effective ways to trigger relaxation when overwhelmed by stress:

- **Control your breathing:** Controlled breathing works by engaging the parasympathetic nervous system, counteracting the body's natural fight-or-flight response induced by stress. By consciously slowing down your breath, you activate a state of calmness and reduce stress levels almost immediately. One effective method is the 4-7-8 technique. Here's how you can practice it:
    - Inhale deeply through your nose for a count of four.
    - Hold your breath for seven counts.
    - Exhale slowly through your mouth for eight counts.
    - This practice not only lowers your heart rate but also stabilizes blood pressure, allowing you a moment to pause and regain clarity during stressful events, ensuring that panic does not impede your decision-making.
- **Practice visualization exercises:** Another beneficial technique for retaining mental stability in crises is visualization strategies. These involve mentally rehearsing successful outcomes, thereby preparing your mind to handle challenges with confidence. Visualization aids in crafting a positive mental framework by encouraging you to imagine yourself navigating potential obstacles with successful outcomes. This strategy is widely used by athletes and military personnel who mentally rehearse their actions before executing them in reality. By picturing yourself remaining calm and in control during a hypothetical emergency, you build resilience and enhance your ability to react effectively when a real crisis arises, knowing you have already *seen* a successful outcome in your mind.
- **Prioritize physical movement:** Physical movement is another key area that can substantially aid in reducing anxiety and tension associated with high-pressure scenarios. Engaging in even mild physical activity triggers the release of endorphins, hormones known for their mood-enhancing properties. Exercises such as walking, stretching, or even light jogging can promote a sense of focus and calmness. Physical movement also helps in managing excess adrenaline, a common

physiological response to stress, which, if unaddressed, can lead to feelings of irritability and panic. By developing an active lifestyle, you can maintain emotional balance, ensuring that stress does not dominate your reasoning capabilities.

- **Use grounding techniques:** Grounding and centering techniques function similarly by keeping you anchored in the present, which is crucial in preventing overwhelming thoughts from taking control during a crisis. Grounding exercises involve redirecting attention away from distressing thoughts toward sensations that root you in the immediate environment. One practical exercise is the *5-4-3-2-1* technique: Identify five things you can see, four you can touch, three you can hear, two you can smell, and one you can taste. This mindfulness exercise promotes heightened awareness of the surroundings and interrupts the spiral of negative thinking.
- **Center yourself regularly:** Centering, on the other hand, focuses energy within yourself to provide stability. It is about finding a personal focal point that brings peace amidst chaos. Techniques like placing your hand over your heart and feeling your heartbeat or reciting a calming mantra can help center your thoughts and emotions. These practices prevent panic-driven reactions, ensuring individuals remain composed and focused on resolving the ongoing situation rather than succumbing to fear or anxiety.

Implementing these strategies requires regular practice and adaptation to individual preferences. It's about finding what resonates best with you and incorporating it into daily routines so that when crises occur, these techniques become second nature. Consistently practicing controlled breathing, visualization, physical movement, and grounding ensures you have a well-rounded toolkit at your disposal, ready to be deployed whenever needed.

## Mindfulness and Mental Training Exercises

Unpredictable circumstances can strike at any moment. From natural disasters to personal crises, being mentally prepared plays a crucial role in how effectively we cope. One powerful approach to bolstering this mental preparedness is through mindfulness practices. The value of mindfulness lies in its ability to enhance clarity and resilience by fostering a heightened state of awareness and presence.

At its core, mindfulness encourages us to focus on the present moment, reducing impulsive reactions that often arise from stress or panic. Imagine being in a situation where every decision you make has significant consequences. In such scenarios, emotional regulation becomes vital. Studies have shown that mindfulness reduces amygdala activity, associated with our fight-or-flight response, creating a space for considered reflection rather than knee-jerk reactions (Schuman-Olivier et al., 2020). This capacity to modulate emotional responses contributes to more measured and rational decision-making processes. Now that you know a bit more about mindfulness, this is how you can go about practicing it:

- **Daily meditation:** A fundamental aspect of mindfulness practice is daily meditation, which helps individuals maintain focus amidst chaos. Meditation trains the mind to return to a point of concentration, even when distractions abound. Breathing exercises are particularly beneficial, acting as an anchor that centers us when chaos ensues. Techniques like box breathing—where inhalation, breath holding, exhalation, and pausing are each counted to four—help integrate this practice into daily life (*Mindfulness STOP Skill*, n.d.). Such exercises can be a reliable tool during high-stress situations, ensuring that panic doesn't cloud judgment.
- **Journaling:** Additionally, journaling emerges as a practical mindfulness tool. It serves as a reflective exercise, allowing individuals to process their emotions, thoughts, and experiences. During times of crisis, writing down feelings can help illuminate patterns in thought and behavior, offering insights into areas needing attention. Before a distressing event, journaling can foster anticipation and strategy development. During the event, it provides a release valve for emotion, and postcrisis, it assists in reviewing what worked well and what didn't, thereby strengthening future resilience.
- **Affirmations and visualizations:** Building a mental toolkit with affirmations and visualization forms another layer of preparation. Affirmations—positive statements reinforcing one's abilities and strengths—build confidence and reduce anxiety levels during challenging times. They remind individuals of their capability to face whatever comes their way. Visualization, on the other hand, involves mentally rehearsing potentially stressful scenarios and imagining effective responses. This practice creates a sense of familiarity with crisis dynamics, reducing shock and improving response readiness when faced with actual pressure.

To implement these practices effectively, it may be useful to follow some guidelines such as the following:

- **Guided mindfulness:** For guided mindfulness practices, setting up a quiet, uninterrupted environment is essential. Whether engaging in meditation or practicing mindful breathing, consistency over time builds the necessary skill set. Moreover, utilize online resources or mobile apps designed to steer beginners gently into the practice.
- **Journaling:** When using journaling as a mindfulness tool, create a structured routine. Dedicate specific times of the day to writing, allowing thoughts to flow without self-censorship. Begin with prompts that explore current emotional states or

hypothetical crisis responses. Over time, this habit will serve as a repository of strategies and reaffirm your growth in handling emergencies.
- **Affirmations and visualization:** To build a mental toolkit, start with simple affirmations tailored to personal fears or insecurities. Repeating these regularly, especially during meditative or mindful moments, ingrains them in the subconscious. With visualization, incorporate vivid imagery that aligns with both your aspirations and the challenges you imagine facing. Practicing visualization consistently familiarizes the brain with these crisis scenarios, paving the way for calmer responses in real-life situations.

## Decision-Making Strategies in High-Stress Scenarios

Being able to make good and logical decisions when emergencies occur is important. High-stress situations can often cloud judgment and create anxiety, making it essential for you to have strategies that enhance your decision-making capabilities. To help you make better decisions, consider the OODA Loop framework. This is a concept developed by military strategist John Boyd, which stands for the following:

- **Observe:** The first step, *Observe*, involves gathering as much current information as possible about the unfolding situation. This might include paying attention to changing weather conditions during a natural disaster or assessing the immediate needs of your family members in an emergency.
- **Orient:** Following this, *Orient* is crucial as it entails analyzing the observed data to form a mental model of the situation. During this stage, you must synthesize new information with your existing knowledge and experiences, allowing you to interpret the scenario accurately and objectively.
- **Decide:** Once a clear understanding is developed, *Decide* occurs, where a course of action is chosen based on the mental model formed. This decision should be both swift and informed, keeping in mind the potential outcomes and consequences.
- **Act:** Lastly, *Act* involves implementing the chosen course of action while remaining vigilant for feedback from the environment.

The OODA Loop is iterative, meaning that after acting, one must return to observing the new situation to adapt and adjust as necessary. By practicing the OODA Loop, individuals can improve their speed and adaptability in decision-making processes, key components of resilience during crises.

## Final Insights

Cultivating mental resilience and preparedness is essential for navigating crisis situations effectively. This chapter explored various strategies to help you and your family cope with stress, emphasizing the importance of understanding stress responses and emotions during emergencies. Recognizing physical and emotional signs of stress can prevent panic and improve decision-making, as you learn to employ calming techniques like deep breathing and grounding exercises. By breaking problems into smaller parts and recalling past experiences, you can make thoughtful decisions even under pressure. Precrisis preparation, through rehearsing emergency plans and fostering open family discussions, equips everyone with knowledge and tools to adapt quickly when unexpected events occur.

The significance of building strong group dynamics and utilizing community support networks was also discussed. Trust and communication within families or groups enhance emotional resilience and ensure everyone's contributions are valued. In this context, regular activities like family meals or game nights reinforce connections, creating a supportive environment. Community support networks further extend this safety net by providing shared resources and collective aid, offering diverse coping strategies and emotional assistance beyond immediate circles. Ultimately, nurturing these relationships ensures that you are all better equipped to face future adversities with confidence, embracing high-quality interactions that promote overall well-being. Once you're mentally prepared for any circumstance, you can move on to some more advanced techniques. In the next chapter, we'll be uncovering some military techniques that you can apply in your own home.

# 9

# LEVERAGING MILITARY TACTICS FOR HOME DEFENSE

You've learned about a number of home defense strategies throughout this book, but let's move the focus to military tactics and their effectiveness. Leveraging military tactics for home defense involves transforming traditional warfare strategies into practical measures that homeowners can employ to enhance the security of their living spaces. This approach is not about turning homes into fortresses but about using tried-and-tested techniques that increase safety and preparedness in a civilian context. The application of these tactics can transform how you perceive and approach home protection, offering fresh perspectives on potential vulnerabilities while providing innovative solutions to reinforce defenses. The goal is to weave principles from established military protocols into everyday settings, creating a proactive stance against threats.

In this chapter, you will learn a range of topics aimed at enhancing home security through the lens of military tactics. Beginning with camouflage and concealment strategies, the discussion will explore integrating natural elements to blend in effectively. The chapter will also cover indoor and outdoor strategies, emphasizing the importance of harmonizing aesthetics with functional security. In addition, we will outline the importance of routine drills and practicing emergency protocols to ensure readiness in real scenarios. Finally, you will gain an understanding of psychological deterrents such as decoys and distractions, which can mislead potential intruders and strengthen overall protective measures.

Camouflage and Concealment Techniques

To effectively obscure a home from potential threats, you can borrow strategies and techniques from military tactics that focus on camouflage and concealment. Understanding natural camouflage is crucial when looking into these military opportunities. Concealing your home is also important, as it has been discussed in this book when talking about securing your home.

- **Camouflage the exterior of your home:** To camouflage your home, it's valuable to use elements of nature to blend in with the structures in your surroundings. For instance, planting shrubs or trees strategically around vulnerable areas, such as windows and entry points, can be highly effective. These natural barriers offer an aesthetic enhancement while serving as a shield against prying eyes. Consider how the shifting seasons might help; dense foliage in spring and summer can act as a screen, while autumn's fallen leaves cover pathways.
- **Conceal the indoors:** Transitioning indoors, creating indoor concealment involves intentionally blending security features into your interior design. Interior spaces should balance between aesthetic appeal and functionality, ensuring safety without making it conspicuous. For example, a sleek bookshelf might hide a panic room door, or furniture could be arranged to obscure surveillance equipment. The idea is to integrate these elements so harmoniously that they remain unseen unless actively sought out. A well-placed mirror, for instance, could deflect attention from reinforced walls or camouflaged safe boxes.
- **Frequent landscaping:** Effective landscaping can enhance camouflage significantly. Using hedges to shield windows and driveways distracts and confuses potential observers. In this setting, it's important to choose plants that match local flora, ensuring they don't stand out. Additionally, climbing plants like ivy can make walls less noticeable and add an extra layer of concealment. Moreover, incorporating rocks, garden paths, and water features can create natural distractions, guiding an observer's eyes away from critical access points.
- **Strategic placements:** Strategic placement of outdoor decorations like sculptures or garden pieces can serve dual purposes: aesthetic enrichment and distraction. A well-placed statue or flower bed draws attention away from certain areas of the property. It's similar to drawing enemy fire away from troop movement in military terms; these ornaments act as visual decoys, focusing attention elsewhere.

We've discussed some of these strategies in greater depth throughout the book, but as you practice them, remember to prioritize a camouflaged and concealed space.

Set up Control Points

When you're going through an emergency, having control points within your home enhances your safety and preparedness. The goal of control zones is to identify key areas that provide strategic advantages in terms of surveillance and communication. These control points serve as vital hubs for managing the flow of information and ensuring a quick response to any potential threats. Here's how you can go about setting up these control points:

- **Identify the optimal control points:** Firstly, identifying control points within your home can be done by leveraging areas that naturally lend themselves to improved surveillance. Look for locations with clear sightlines or those offering natural cover, such as behind thick curtains or tall furniture. These spots should enable you to monitor key entry and exit points without exposing yourself unnecessarily. For instance, a window in a high-traffic area might allow for discreet observation while maintaining a safe distance from potential threats. Using these vantage points, you can gather crucial information about the outside environment, assessing risks before they become immediate dangers.
- **Customize these control points:** Customizing control points to fit your personal security needs is another essential step. Utilize existing furniture creatively to facilitate monitoring. A bookshelf or wardrobe, for instance, can be repositioned to distract attention from critical areas of your home, allowing you to observe incoming paths without being seen. Additionally, incorporating strategic maps of your home helps family members understand escape routes and safety zones, ensuring everyone knows where to go in case of an emergency. This personalized setup provides a structured approach to managing crises, empowering you to respond effectively under pressure.
- **Establish communication guidelines:** Establishing effective communication protocols is important for coordinating efforts during a crisis. To achieve this, you should set up systems that allow all household members to efficiently relay information, even through nonverbal cues. Depending on the severity of the situation, verbal communication may not always be possible. In these cases, implementing a series of hand signals or flashlight codes can act as silent alerts between family members. Regular drills and practice sessions can help reinforce these communication methods, ensuring everyone is familiar with their meaning and application.
- **Secure your control points:** Reinforcing control points with secure locks, barriers, and essential supplies further enhances their functionality during emergencies. Installing heavy-duty locks on doors and windows at these critical locations ensures intruders have difficulty breaching them. Barricades, such as furniture or precut wooden planks, can offer additional reinforcement against forced entries. Remember that equipping these areas with necessary supplies, including first aid kits, emergency contact lists, and basic tools, prepares you for various scenarios. A well-stocked control point acts as a safe haven, providing both physical protection and the resources needed to sustain communication and surveillance for extended periods.

Consider a typical scenario where these measures could be put to use. Imagine a power outage in your neighborhood that leads to increased nighttime burglaries. Your identified control points, perhaps near a front-facing window and a back door, allow you to maintain a watchful eye on suspicious activities without revealing your presence. Thanks to prearranged furniture setups, you can quickly ascertain the situation outside while remaining comfortably hidden. Communication protocols enable you to silently alert other family members using predetermined signals, ensuring everyone stays informed and ready to act if necessary. With reinforced barriers and crucial supplies within reach, you are well-equipped to handle prolonged disruptions until normalcy resumes.

Use *Fieldcraft* Skills

Using military tactics for home defense requires the use of practical *fieldcraft* skills. These skills can greatly enhance the security of your household, providing an edge in navigating potential threats with efficiency and precision. Here are some *fieldcraft* skills you should consider:

- **Silent movement techniques:** One of the foundational elements in adapting military tactics for home defense is the establishment of silent movement techniques. In urban settings, noise can be both a distraction and a hazard. Learning how to move quietly is imperative for maintaining the element of surprise and avoiding detection by intruders. This involves altering standard movements to reduce sound production, such as practicing soft footfalls, mindful breathing, and minimizing contact with noise-producing surfaces. For instance, practicing a heel-to-toe walking pattern can significantly lower the noise generated while moving through a house. Additionally, wearing soft-soled shoes generates less noise compared to hard-soled footwear. Mastering these techniques enables you to navigate your home stealthily during emergencies, making it more difficult for potential threats to detect your presence.
- **Situational awareness:** Situational awareness is another critical component of home defense, enhancing decision-making capabilities under pressure. This skill entails consistently scanning your surroundings and processing environmental cues to form mental maps. Mental mapping allows for rapid evaluation of potential threats and aids in quick strategizing in high-stress situations. This situational awareness allows you to preemptively identify anomalies or suspicious activities, potentially deterring a threat before it fully manifests. An easy way to practice this is by mentally noting exits, potential hiding spots, and choke points in familiar spaces. Developing a routine of regularly assessing changes in the environment strengthens this skill, creating a more secure and vigilant domicile.
- **Adaptable survival techniques:** Beyond immediate physical defense strategies, adaptable survival techniques play a crucial role in sustaining safety during urban

incidents. Situations such as natural disasters or prolonged civil unrest require residents to possess basic skills like emergency first aid and quick repair methodologies. Being adept at performing urgent fixes to broken windows, damaged locks, or malfunctioning utilities can prevent vulnerabilities that intruders might exploit. In addition, having knowledge and supplies for basic medical interventions ensures preparedness for injuries sustained during crises. Along with your well-stocked first aid kit, you should learn simple procedures such as wound dressing, CPR, and using a tourniquet. Such preparations empower you to maintain control over your environment, reducing dependency on external assistance in adverse situations.

- **Regular drills:** To further enforce these skills, structured practice and regular drills are beneficial. Consistent training enhances proficiency and boosts confidence, ensuring that responses to threats are second nature. Simple drills, like timed exit scenarios or silent movement challenges, provide practical experience and reveal gaps in existing plans. Encouraging active participation reinforces commitment to personal and family safety, building a resilient community mindset.

When integrating military-derived tactics for home defense, it's important to always prioritize safety and legality. Many techniques are initially designed for combative contexts and may need adaptation to suit civilian applications without escalating risks. Consulting with security professionals or participating in community safety workshops can provide valuable insights and avoid potential missteps.

### Implement Patrols and Surveillance

Incorporating military tactics for home defense can significantly improve a household's safety and preparedness. By establishing methods for regular patrols and surveillance, you can proactively monitor your environment, deter potential threats, and respond effectively to emergencies. This approach enhances individual home security and also encourages community collaboration, creating a safer neighborhood. Here are some steps that can help you implement patrols and surveillance:

- **Maintain continuous vigilance:** To start, creating a patrol schedule is essential for maintaining continuous vigilance. It involves designing a rotating system among family members, ensuring that there is always someone responsible for monitoring the home. This system is similar to military watch schedules, where duties are divided among personnel to cover shifts throughout the day and night. The key to an effective schedule lies in its consistency and adaptability. Family members should be trained to recognize unusual activities and communicate clearly with each other when passing responsibility from one person to another. An efficient rotation does more than just maintain vigilance; it allows each family member the chance to develop observational skills and situational awareness.

- **Use surveillance equipment:** Utilizing surveillance equipment further fortifies a home's defensive capabilities. Today, we are fortunate enough that technology offers numerous options, ranging from high-tech security cameras to simpler, low-tech solutions like noise sensors. Security cameras provide a visual record and can be strategically placed to monitor entry points or vulnerable areas around the home. Many modern systems offer remote access, allowing homeowners to monitor their property even when they are away. However, it's crucial not to overlook the value of low-tech devices. Noise sensors, for instance, can provide alerts about unusual sounds such as shattering glass or forced entry. These tools become the eyes and ears of the household, enhancing overall awareness and offering critical early warnings of potential intrusions.
- **Report protocols:** An additional layer of security comes from implementing helpful reporting protocols. Establishing a basic log system for family members helps document observations and any unusual incidents. This habit influences an organized approach to tracking potential threats and patterns over time. Keeping detailed notes on dates, times, descriptions of suspicious activities, and actions taken can be extremely beneficial for law enforcement during investigations. Furthermore, encouraging consistent reporting improves each family member's ability to quickly assess situations and relay accurate information, which is vital during emergencies. Training sessions that combine practical scenarios and support family preparedness reinforce these principles by simulating real-life situations.

Again, regular drills, including children, can make these patrols and surveillance more effective when in action.

## Utilizing Decoys and Distractions

Using decoys and distractions is one effective way to strategically and proactively secure your home. Understanding the use of these decoys and distractions as part of your home defense strategy can be invaluable in misleading potential threats and fortifying your safety measures.

The concept of using decoys as a defensive tactic derives from the principle of misdirection—creating illusions that confuse or divert an intruder's focus away from actual vulnerabilities. Imagine a scenario where your home appears equipped with multiple cameras and reinforced locks on every door. While some of these may indeed function as security devices, others might be cleverly disguised as nonoperational or decoy versions. This technique creates a false sense of security for any would-be intruder, causing them to second-guess whether an entry point is truly accessible. The following decoys and distractions can be quite effective:

- **Visual decoys:** Decoys such as fake surveillance cameras or alarm signs can act as deterrents, potentially persuading an intruder to move on to what they perceive as an easier target.
- **Audible decoys:** Audio devices that simulate human activity, like the noise of a barking dog or the chatter of a television, can create the illusion of occupancy even when no one is home.
- **Timers:** Consider setting these sounds on timers or activating them remotely to maintain the facade of a lively household. The unexpected nature of these noises serves as a psychological barrier, sowing doubt and hesitation in an intruder's mind.

Implementing these tactics requires careful planning and continuous reassessment. These steps can help you make the most of these decoys:

1. **Evaluate the perfect spots:** First, evaluate the most strategic locations around your home to place decoys and sound generators. Entry points such as doors and windows are obvious choices, but consider less apparent areas that might offer cover to an intruder.
2. **Practice regular assessments:** Once set up, regularly assess their effectiveness. This might involve checking that decoy cameras remain visible yet unobtrusive and ensuring audio devices blend naturally into daily routines without raising suspicion among neighbors. Routine testing and drills are critical components in maintaining a robust defense system. Hence, conduct regular walk-throughs of your home's layout to identify potential weaknesses or blind spots that an intruder might exploit.
3. **Be Adaptable:** As with any military operation, strategies must evolve over time to counter new tactics utilized by potential intruders. Update your decoy setups periodically to keep them credible, and research emerging technologies that could enhance your current system. Introducing new sound devices that mimic ongoing conversations or footsteps, for instance, can add an extra layer of unpredictability, keeping your defenses one step ahead.

Final Thoughts

By applying military tactics to enhance home defensive capabilities, you can create a more secure environment for your family. The chapter discussed some camouflage and concealment techniques, illustrating how blending natural and artificial elements can obscure your home from unwelcome attention. As we moved indoors, we showcased the importance of integrating security features seamlessly into the design without compromising aesthetics. The strategic use of technology further fortifies these efforts, simulating occupancy and offering discreet surveillance. Routine drills empower family members with familiarity with emergency protocols, ensuring swift action is possible when needed.

Setting up strategic control points helps manage crises effectively through improved surveillance and communication. These hubs act as command centers, fitted with necessary tools and barriers to withstand potential threats. Incorporating patrols and surveillance into daily routines maintains constant vigilance, while community collaboration enhances security measures beyond individual households. The use of decoys and distractions adds another layer of defense by confusing potential intruders. Together, these strategies create a disciplined approach that arms homeowners with the knowledge and skills required to safeguard their living spaces. With all of these military techniques in place, you should consider gathering community support next.

# 10

## BUILDING A SUPPORTIVE COMMUNITY NETWORK

A community of support can drastically boost your preparedness when faced with challenging crises. Building a supportive community network involves nurturing relationships with neighbors, which can significantly enhance collective security. When you and your local community collaborate toward a shared goal, you contribute to your own safety and also to that of the entire neighborhood.

Let's take a look at how you can foster these relationships within your local area, which leads to a strong security framework. By understanding this sense of interconnectedness, you can transform isolated efforts into cohesive strategies that benefit everyone. The underlying concept is simple: When people come together with a common purpose, their combined vigilance and care create an environment where crime becomes less likely, and support is readily available during emergencies.

In this chapter, we will explore various practical approaches to building this network, focusing on neighborhood watch programs as a primary strategy. It highlights the importance of community participation in reporting suspicious activities and the role of local leaders in streamlining communication between residents and law enforcement. The chapter also discusses the integration of modern technology to enhance surveillance and communication within the community, ensuring all members are connected and informed. By continuously evaluating and adapting these strategies, your community can remain prepared to tackle emerging threats effectively.

## Benefits of Community Watch Programs

Community watch programs are initiatives where members of a neighborhood come together to watch over their community. These programs encourage residents to take an active role in the safety and security of their environment.

### *Sense of Security*

One of the main benefits of these programs is the increased sense of security that comes from people being vigilant. When neighbors are aware of what's happening around them, they can notice any suspicious activity and report it quickly. This awareness can deter crime because potential offenders may think twice if they know the community is watching.

### *Sense of Community*

Another significant benefit is the sense of community that develops through these programs. People who participate in community watch programs often get to know their neighbors better. They communicate and build relationships, which helps foster a sense of belonging. For example, regular meetings or gatherings can be organized to discuss safety concerns, plan community events, or simply enjoy each other's company. This social interaction can strengthen the bonds between residents, making the entire neighborhood feel more like a home.

### *Education on Safety*

Community watch programs also provide education on safety and crime prevention. Residents can learn about common crime tactics and how to protect themselves and their property. Workshops can cover a wide range of topics, including home security tips and emergency response procedures. For instance, communities might organize a workshop where residents learn how to secure their doors and windows effectively. Participants can also be educated on recognizing suspicious behavior and how to report it to the proper authorities.

### *Collaboration With Law Enforcement*

Additionally, these programs can work in collaboration with local law enforcement. When community members team up with police officers, they can communicate more effectively about concerns and trends in criminal activity. Officers can share information about crime statistics and offer advice on prevention. This partnership fosters a better relationship between residents and the police, helping to create a cooperative environment. For example, some communities have regular meetings with local officers to discuss recent changes in crime patterns, which can empower residents and keep them informed.

### *Engaging the Younger Generation*

Engaging the younger generation is another benefit. Many community watch programs involve children and teenagers, teaching them the importance of being responsible citizens. These young people can learn valuable lessons about safety and vigilance while developing a sense of duty toward keeping their community safe. Activities can include youth-led initiatives like poster campaigns or peer education sessions on safety, helping to instill responsibility and community care from a young age.

### *Encouraging Communication*

Moreover, community watch programs encourage communication among residents. Having a structured way to communicate can be very beneficial. Many neighborhoods create newsletters or use social media groups to share information about safety tips, upcoming meetings, or local events. This regular communication keeps everyone informed and involved, fostering a proactive mindset. For instance, if a resident notices an unusual vehicle in the neighborhood, they can easily share this information through their community's platform, allowing others to stay alert.

## Creating Community Watch Programs

Creating a safer community through neighborhood watch programs encourages more vigilant participation from all residents. If you are interested in starting community watch programs and you think your neighborhood would be as well, consider the following tips:

- **Create community awareness:** Establishing awareness among community members begins by encouraging everyone in the community to report any suspicious activity. This simple yet effective act transforms ordinary citizens into proactive guardians of their local area. When neighbors keep an eye out for unusual behavior, it deters potential criminals and influences a culture of mutual care and responsibility. This collective vigilance creates a safer environment as everyone contributes to the overall security of the community, leading to a sense of shared purpose and protection.
- **Engage with local leaders:** Central to the success of any neighborhood watch program is the engagement of local leaders and the establishment of structured roles. Involving respected figures within the community, such as homeowners association heads or civic group leaders, naturally enhances communication. These leaders serve as crucial liaisons between the community and law enforcement agencies. When these structured roles clearly outline responsibilities—such as coordinators, street captains, and designated communicators—the flow of information becomes efficient and transparent. The presence of a defined hierarchy streamlines operations and also ensures that there is always someone accountable for each aspect of the program. A system like this makes it easier to organize meetings,

release relevant updates, and coordinate actions during emergencies, thereby strengthening the program's overall effectiveness.
- **Host training sessions:** Training sessions are an equally important part of empowering community members to contribute effectively to crime prevention. These sessions cover aspects of personal safety, emergency handling, and preventative measures against common crimes like theft or vandalism. When community members participate in training sessions, they're able to identify telltale signs of criminal activity, which enhances their observational skills and confidence in distinguishing normal occurrences from potential threats. These sessions also often include collaboration with local law enforcement, who can provide valuable insights into crime trends and offer practical advice on maintaining neighborhood safety.
- **Educate your community:** In addition to general crime prevention training, more specialized education can be beneficial. For instance, understanding how to use personal defense tools or mastering basic first aid can be invaluable during certain emergencies, especially for people in the community who are unaware of how to operate different tools. Appropriate training prepares everyone with the necessary skills to act decisively should a crisis occur. It also contributes to a greater sense of empowerment and responsibility within the community, promoting a self-sufficient mindset where all community members feel capable of managing situations independently before professional help arrives.
- **Share experiences with each other:** While structured training is crucial, informal knowledge-sharing among residents should not be underestimated. Organizing small groups or workshops where experiences and tips are discussed can be incredibly enriching. These gatherings pave the way for the exchange of creative ideas on improving home security, such as using smart technology for surveillance or implementing neighborhood patrols during vulnerable times. These interactions also reinforce community bonds, making people more willing to participate actively in keeping their surroundings safe.
- **Use technology to communicate:** Communication technology also helps maintain real-time contact among community members. Online platforms like dedicated social media groups or community forums allow for the swift sharing of information, alerts, and updates. These platforms can host detailed maps indicating recent suspicious activities, thus providing residents with a comprehensive view of potential risks. Moreover, using messaging apps for direct communication ensures that alerts reach all members quickly, enabling coordinated actions and swift reporting of incidents to authorities.
- **Collaborate with law enforcement:** Collaborating with law enforcement to conduct periodic audits or assessments of the watch program can offer valuable insights into its effectiveness. These evaluations include processes such as reviewing incident reports, analyzing patterns, and pinpointing weaknesses in existing

protocols. Implementing recommendations from these audits reinforces the watch program's resilience and reliability.

It's crucial for neighborhood watch programs to look beyond crime prevention and consider broader safety concerns affecting the community. Engaging in initiatives like community cleanup events not only deters graffiti and vandalism but also improves the aesthetic and psychological environment of the neighborhood. Clean, well-maintained areas are less attractive to criminals and more inviting to residents, which contributes to a sense of pride and belonging.

### Establishing Mutual Aid Agreements

Developing mutually beneficial relationships can provide you with aid during times of crisis when you need it most. Establishing mutual aid agreements plays a crucial role in resource sharing and support among neighbors, creating a sense of cooperation, resilience, and shared responsibility.

Mutual aid is defined as the process where two or more parties voluntarily help each other with resources or skills during emergencies. This practice not only strengthens community bonds but also increases collective security by bringing individuals together for a common purpose. Within such frameworks, neighbors aren't just isolated units; they function as part of a broader safety net that thrives on collaboration. These steps can help you create mutual aid agreements:

1. **Conduct a thorough resource inventory:** A fundamental step in implementing effective mutual aid is conducting a thorough resource inventory. This can be achieved by cataloging the available resources within your community—from food supplies and medical equipment to generators and transportation means. Identifying what is at hand can help communities ensure efficient emergency responses without leaving anyone vulnerable. Imagine a sudden power outage; if one home has extra generators and another has an abundance of nonperishable foods, mutual aid agreements can facilitate quick exchanges to benefit all parties involved. This systematic approach ensures no household feels left out or overwhelmed during times of need.
2. **Organize regular meetings:** Creating these inventories requires active participation from the community. Residents can organize meetings to discuss individual contributions and collectively plan how best to utilize shared items. During these gatherings, it's beneficial to appoint coordinators who maintain records of available resources and monitor changes in supplies over time. This proactive measure guarantees that if disaster strikes, communities are prepared and capable of handling challenges through concerted efforts.

3. **Draft clear guidelines:** Beyond identifying resources, drafting clear guidelines for aid is essential for smooth operation during crises. These guidelines function as contracts that solidify commitments and manage expectations. They provide clarity on roles, responsibilities, and procedures, ensuring everyone knows what to do and whom to contact when assistance is needed. For instance, clearly outlining protocols on how aid requests should be communicated—whether via phone, group texts, or digital platforms—prevents confusion and delays in response times. When creating these guidelines, it is important to involve all participants in discussions to accommodate diverse needs and capacities. Understanding each member's strengths and limitations empowers the community to strategize effectively. Furthermore, regularly revisiting and updating these guidelines keeps them relevant and reflective of any changes within the community's dynamics or resources.

Such organization within mutual aid agreements mirrors systems already present at larger scales. Formal agreements, like those mentioned by the National Incident Management System (NIMS), set a legal basis for entities to share resources seamlessly across jurisdictions, including between neighboring communities or even states (*Performance Under Pressure*, 2017). While communities may operate on a smaller scale, adopting similar principles enhances preparedness and operational efficiency.

Beyond practical aspects, mutual aid agreements fundamentally instill resilience. Neighbors participating in resource-sharing arrangements often develop stronger social ties, which further enhance community morale. The knowledge that there is a network of supportive individuals willing to lend a hand can reduce feelings of isolation and anxiety during daunting situations.

These agreements also encourage skill-sharing, which is particularly useful in emergencies requiring specific expertise. Whether it's first aid training, technical know-how on repairing infrastructure, or logistical planning, mobilizing diverse skills within the neighborhood enriches everyone's knowledge and readiness.

Coordinating Neighborhood Emergency Plans

We've discussed how you can create emergency plans for yourself and your family, but let's extend it to the community. Creating a unified emergency plan is essential for ensuring a community's safety and preparedness. One of the primary objectives when building a supportive community network is to consolidate efforts into a cohesive strategy that encompasses everyone involved. Doing this encourages better coordination and an enhanced understanding of procedures that can save lives during emergencies.

Developing such a plan requires inclusive participation from all members of the community. It's important to establish an environment where everyone feels empowered to contribute their ideas and experiences. Furthermore, it's crucial to involve individuals from diverse backgrounds and with varied expertise so that the plan becomes more adaptable to different situations.

Everyone needs to be aware of their roles and responsibilities during an emergency to prevent chaos and confusion. Clear communication about who does what, whether it's shutting down utilities or attending to neighbors in need, aids in executing the plan smoothly when time is of the essence.

Mapping key resources within the community is another critical component of effective emergency planning. Accurate maps displaying essential locations like safe shelters, hospitals, and emergency supply depots allow families to quickly navigate to necessary services when incidents occur. These maps should be accessible both physically—with printed copies distributed among residents—and digitally, as many people rely on smartphones for real-time updates. Incorporating local landmarks makes these maps intuitive, especially in high-stress situations where clear thinking might be compromised. Consideration should also be given to regularly updating these resource maps to reflect any changes in infrastructure or service availability, ensuring their reliability when they are most needed.

As part of these educational initiatives, residents should be encouraged to share personal experiences and strategies related to past emergencies. Learning from real-world situations can provide invaluable insights into what works and what doesn't. Additionally, inviting experts to speak at community gatherings offers fresh perspectives and introduces new techniques that may improve existing plans.

Sharing Resources and Skills

Creating a supportive community is great because it helps us share resources and skills. By doing this, communities can strengthen their ability to face emergencies and everyday challenges. Let's take a look at some ways sharing resources and skills as a community can help during emergencies:

- **Identifying the skills everyone has:** Understanding the distinctive skills that each community member brings to the table is essential in enhancing self-sufficiency during emergencies. Every individual possesses talents or knowledge that can be invaluable when crises arise. For instance, someone with medical training can provide first aid during emergencies, while another with technical skills might help restore power or communication lines if they fail. Organizing skill-assessment workshops or surveys can help identify these abilities effectively. Community leaders can then compile this data into a directory, making it easier to mobilize appropriate resources quickly when needed. This approach creates a sense of belonging while promoting confidence within the community, knowing that they have internal capabilities to rely on.

- **Pooling resources together:** Pooling resources is another vital strategy for strengthening community resilience. By centralizing supplies such as food, water, and emergency tools, communities can create a collective safety net that ensures everyone has access to necessities during critical times. This practice reduces the financial burden on individual families and promotes mutual aid, where neighbors support one another by sharing surplus or specialized items. Establishing community storage facilities or supply depots can be an effective way to manage these resources. With clear guidelines on contribution and distribution, the system can function smoothly, providing much-needed relief and security to all residents.
- **Integrating local and traditional practices:** Integrating traditional practices and local knowledge adds depth to community resilience efforts. Many communities possess cultural heritage and ancestral wisdom related to survival and resource management. Recognizing and incorporating these practices can enrich contemporary strategies, making them more flexible. Moreover, involving elders and long-term residents in planning and decision-making processes can bridge generational gaps and ensure diverse perspectives are considered.
- **Proactive leadership:** A successful community network relies on proactive leadership to steer initiatives and inspire participation. Leaders can emerge from any part of the community, and it is their enthusiasm and dedication that drive collective efforts forward. Effective leaders listen to their community's needs, create inclusion, and delegate tasks according to members' strengths. Leadership training programs can be instrumental in nurturing potential leaders and equipping them with the skills necessary to guide their communities through challenging times.

As community dynamics change and challenges evolve, the strategies you choose collectively must remain flexible. Regular feedback loops—whether through surveys, meetings, or digital channels—help communities assess the effectiveness of their initiatives and identify areas for improvement. This process ensures that the community remains resilient and ready to face new adversities with agility.

## Hosting Regular Safety Meetings

Engaging families through regular safety meetings is an essential strategy to enhance community participation and build a robust support network. Families can be informed participants in their neighborhood's security by cultivating a culture of continuous dialogue. Establishing the frequency of these meetings can create routine discussions, allowing you to anticipate and prepare for them. Consistency encourages a sense of inclusivity, encouraging diverse perspectives, which enriches discussions and decision-making processes. When everyone feels that their voices matter, it strengthens the bond within the community, creating a safer environment for all. Here's how you can plan these regular safety meetings:

- **Set regular schedules:** Setting a regular schedule for safety meetings allows communities to maintain momentum in their security efforts. Whether monthly or quarterly, consistent gatherings provide a platform for ongoing conversations about local safety issues. These meetings serve as checkpoints to evaluate previous initiatives' progress and assess new objectives. More importantly, they offer an opportunity to welcome new residents and integrate them into the community network effectively. The frequency also ensures no critical concerns fall through the cracks, and it keeps everyone updated on recent developments or threats.
- **Craft agendas:** It's also important to consider crafting agendas to keep meetings focused and productive. A well-structured agenda guides attendees through essential topics while leaving room for open discussion. This structure encourages resident collaboration and topic suggestions, making it clear that everyone's input is valued. Agendas can begin with reviewing past meeting notes and updates on ongoing projects, followed by addressing new business, such as emerging neighborhood concerns or policy changes. Lastly, they should allocate time for brainstorming and proposing future actions, ensuring meetings remain dynamic and solution-oriented.
- **Incorporate flexibility:** An efficient agenda is not cast in stone but remains flexible to accommodate pressing issues and member contributions. For example, if a particular concern, like increased traffic incidents, arises, it can be prioritized on the next meeting's agenda. Residents might also propose topics via email or suggestion boxes before the meeting. These practices reflect transparency and empower residents, reinforcing their role as active participants in maintaining neighborhood safety.
- **Invite expert speakers:** Inviting expert speakers to these safety meetings can substantially elevate the value members derive from attending. Experts provide additional insights and valuable training on health and safety procedures, bridging potential knowledge gaps among residents. Their expertise might span various safety-relevant fields, such as emergency preparedness, crime prevention, or first aid training. Offering practical demonstrations or workshops during meetings equips community members with the skills necessary to respond effectively to emergencies. For instance, a speaker from a local fire department could conduct a session on home fire prevention techniques and emergency evacuation planning. Likewise, a representative from law enforcement might explain the intricacies of reporting suspicious activities, demystifying police processes, and establishing trust between officers and the community.
- **Continuous effort and dedication:** Integrating families into a proactive safety culture demands continuous effort and dedication. With that being said, the rewards are worth it as they manifest in a more cohesive, secure community environment. By sustaining meeting frequency, crafting inclusive agendas, and leveraging expert knowledge, residents take ownership of their collective safety. As families participate

regularly, they become more knowledgeable and connected, translating individual engagement into communal resilience.

Moreover, these gatherings nurture a shared responsibility model where everyone has a part to play. They emphasize that community safety isn't merely the job of authorities but a collaborative effort requiring everyone's involvement. Children, too, learn from observing their parents' participation in civic duties, setting an example of community service and vigilance early on. Ultimately, regular safety meetings bridge communication gaps and strengthen neighborhood bonds. Though logistical challenges such as scheduling conflicts and diverse availability may arise, solutions like rotating days and times or using virtual platforms can mitigate these barriers, ensuring inclusivity.

Reflection

Throughout this chapter, we have explored the importance of creating healthy relationships with neighbors through community watch programs and mutual aid agreements. By engaging everyone in active participation and structured roles, these initiatives create a network of vigilance and support that enhances collective security. Training sessions and informal knowledge-sharing empower you with skills to prevent crime and respond effectively to emergencies as a community. Additionally, integrating modern technology and maintaining open communication channels ensures swift action and real-time updates, further enhancing safety efforts.

In tandem, the chapter emphasizes the significance of establishing mutual aid agreements to strengthen community bonds and resource sharing during crises. Conducting resource inventories and drafting clear guidelines for aid help organize contributions and manage expectations, ensuring a coordinated response when needed. Hosting regular safety meetings further deepens family engagement and enriches discussions with expert insights, promoting a culture of shared responsibility. Together, these approaches create an adaptable and cohesive community environment dedicated to safeguarding its members. With community involvement achieved, it's important to look at ways to adapt to different crises. While you may be prepared for the most likely emergency, having the ability to adapt under pressure can be your saving grace. This will be discussed in the next chapter.

## 11

# ADAPTING TO DIFFERENT CRISIS SCENARIOS

Unfortunately, we can't predict what is going to happen next in our lives, hence the necessity to adapt to different crisis scenarios. Each type of crisis, whether it be natural disasters, civil unrest, health emergencies, or other potentially destabilizing events, presents unique challenges and necessitates tailored solutions. Understanding these distinctions is crucial for developing effective strategies that ensure both immediate safety and long-term resilience. These differences showcase the importance of having a well-rounded preparedness plan that accounts for various types of crises.

In this chapter, we will explore a diverse range of crisis situations and learn how to customize your preparedness strategies accordingly. The discussion will cover the essential elements of creating and implementing plans that address specific threats such as natural disasters and health crises. We'll also be covering how to face long-term power outages and hazardous materials that can be threatening to you and your family's safety. Understanding these varied aspects of crisis adaptation can help you navigate complex situations effectively.

Preparing for Natural Disasters

Natural disasters are some of the often-unpredictable emergencies that can pose a threat to you and your family. Common natural disasters like floods, earthquakes, hurricanes, and wildfires can have devastating impacts on communities worldwide. Each disaster type has specific characteristics that require tailored responses, making it vital for you to be well-informed about the hazards prevalent in your area. Here are some ways you can prepare yourself and your home for different natural disasters:

- **Consider different natural disasters:** Floods can occur due to heavy rains, storm surges, or rapid snow melts. They often lead to property damage, loss of life, and long-term disruptions. Earthquakes, on the other hand, strike without warning, causing buildings and infrastructure to collapse. Understanding the behaviors and triggers of such disasters helps residents anticipate challenges and prepare accordingly.

- **Be aware of your geographic vulnerabilities:** Recognizing geographic vulnerabilities is crucial for effective crisis management. Coastal areas are particularly prone to hurricanes, while certain inland regions face higher flooding risks due to river systems. Earthquake-prone zones, especially those near tectonic plate boundaries, must prioritize seismic resilience. Similarly, arid regions are more susceptible to wildfires, which can spread rapidly under dry conditions. Analyzing these vulnerabilities as a community can help you uniquely prepare for the most pressing environmental threats you face.
- **Secure your home against potential natural disasters:** Securing homes against potential disasters involves a thorough assessment of structural strengths and weaknesses. Retrofitting homes with earthquake-resistant features, such as shear walls and foundation bolting, enhances stability during seismic events. Flood-proofing may include elevating structures, installing sump pumps, and constructing barriers. For regions vulnerable to hurricanes, securing roofs and reinforcing doors and windows with storm shutters can prevent wind damage. Removing hazardous materials and securing heavy furniture and fixtures can further reduce risks during emergencies. These modifications create a safer living environment and increase resilience against natural disasters.
- **Develop family emergency plans:** Developing family emergency plans is an essential component of disaster preparedness. These plans should clearly outline the roles and responsibilities of each family member, ensuring everyone knows what to do in an emergency. Roles might include identifying a primary caregiver for young children, who will gather emergency supplies, and who will manage communication with authorities and neighbors. When crafting a family emergency plan, it's important to consider all scenarios, including evacuation and shelter-in-place strategies. Families should identify safe areas within their homes for taking cover during storms and know the nearest evacuation routes and public shelters.

As always, you can never go wrong with creating clear guidelines when assembling emergency kits; for instance, you should prioritize essential items for survival and comfort.

## Considering Pandemics and Health Crises

We've had a recent pandemic, which left most of us completely unprepared and blindsided.

Keeping yourself prepared with strategic planning can ensure your safety. Here's how you can equip yourself for pandemics and health crises:

- **Identify potential health risks:** The first step in this process involves identifying potential health risks by comprehending regional contagion threats and associated symptoms. This understanding encourages preparedness in actions. For example, regions prone to vector-borne diseases such as malaria or dengue should prioritize measures against mosquito bites, emphasizing the need for protective clothing and insect repellent. Conversely, areas susceptible to respiratory infections like influenza or COVID-19 would benefit from heightened awareness around symptoms like fever, cough, and difficulty breathing, prompting timely medical consultation (Dhaka et al., 2021).
- **Create a health safety plan:** Creating a comprehensive health safety plan is essential in navigating these scenarios effectively. Emphasizing hygiene practices and sanitation plays a crucial role in curbing the spread of infections. Simple yet effective habits such as frequent hand washing with soap, proper disposal of waste, and maintaining clean living spaces can significantly reduce the risk of infection. It's also beneficial to establish designated hand sanitizing stations within households and reinforce these practices regularly, especially among children, to instill long-lasting hygienic behavior.
- **Stock essential health supplies:** Stocking essential health supplies is another foundational component of health emergency preparedness. Focus should be placed on personal protective equipment (PPE) and maintaining reserves of essential medications. Households should assess their needs based on family size and existing health conditions, ensuring they have ample supplies of masks, gloves, disinfectants, and basic first aid items readily available. It's also important to consult with health care providers about stocking up on necessary prescription medications to avoid any interruptions in the management of chronic conditions during emergencies. Regularly checking expiration dates and replenishing stock as needed is also essential for maintaining preparedness.
- **Have virtual medical consultations:** Utilizing technology for virtual medical consultations has emerged as a groundbreaking approach to managing health emergencies efficiently. Telemedicine platforms allow for rapid information sharing between patients and health care providers while minimizing the need for physical office visits (Filip et al., 2022). Particularly in remote or underserved areas, this technological leap can ensure that you and your family receive appropriate medical advice and treatment without delay, reducing the burden on emergency rooms.

Implementing these measures calls for ongoing evaluation and adaptation. As health threats evolve, so should our preparedness strategies, adapting to new challenges posed by advancing pathogens or changing environmental conditions.

## Dealing With Long-Term Power Outages

It's common for unforeseen circumstances to result in power outages that last long. Preparing for potential outages can ensure safety and comfort during these circumstances. This guide can help you prepare and manage these power outage challenges:

- **Identify the causes of the power interruptions:** Understanding the causes of these power interruptions is an essential first step in building resilience. Outages can occur due to infrastructure weaknesses, such as aging power lines or transformers that fail under stress. Weather-related events, including severe storms, hurricanes, or snowfalls, are also common causes; these natural phenomenas can disrupt electrical supply for extended periods. Acknowledging these potential risks allows you to better anticipate and mitigate the impact of power loss.
- **Explore alternative power sources:** To combat the effects of power outages, it is important to explore alternative power sources. Generators are a popular choice for providing backup power capable of running essential household appliances. When selecting a generator, consider the power output needed to sustain basic operations and ensure proper installation by following manufacturer guidelines. Solar panels offer another sustainable option, converting sunlight into electricity. Although initial setup costs can be significant, solar panels offer long-term energy savings and reduce reliance on external power grids. Installing battery storage systems alongside solar panels can further enhance self-sufficiency during outages, storing excess energy generated during daylight hours for use at night or during cloudy weather.
- **Prioritize food safety:** Managing food safety is critical during power outages, as the lack of refrigeration can compromise perishable goods. To mitigate spoilage risks, stock up on nonperishable foods, such as canned goods, dried fruits, nuts, and grains, which do not require refrigeration and have long shelf lives. In this setting, developing meal strategies in advance can help maximize your resources. Plan meals around available ingredients, prioritize consuming refrigerated items before they spoil, and consider using camping stoves or grills for cooking if safe to do so outdoors. The USDA suggests keeping appliance thermometers in refrigerators and freezers to monitor temperatures and assess food safety after power returns. If any food items exceed safe temperature thresholds, they should be discarded to avoid foodborne illness (*Are You and Your Food Prepared for a Power Outage?*, 2021).

Maintaining a comfortable living environment without electricity poses unique challenges. Encourage family engagement through indoor activities like board games, storytelling, or reading, which foster bonding and distract from the absence of electronic entertainment. During daylight hours, outdoor activities such as hiking, gardening, or yard games provide fresh air, exercise, and a sense of normalcy. Adapting to life without electricity may involve creating new routines and adapting old habits. Consider hand-powered tools and gadgets or solar-powered chargers for essential devices.

How to Deal With Hazardous Materials

When dealing with hazardous material incidents, understanding and managing risks is necessary for ensuring safety in both homes and communities. These are the steps you should follow to manage hazardous materials:

- **Recognize potential hazards:** Recognizing potential hazards such as household chemical cleaners and nearby industrial activities is the first step toward mitigating threats posed by hazardous materials. These substances include chemicals that can be explosive, flammable, or toxic. Their mishandling during production, storage, or disposal can result in perilous situations. It's important to familiarize yourself with the types of chemicals typically used at home, as well as their proper storage methods to prevent accidental exposure. For instance, keeping cleaning supplies in their original containers and away from children or pets reduces the risk of accidental ingestion or contact.
- **Utilize community awareness:** Again, community awareness of industrial activities can help residents prepare for any potential emergencies. Industries often hold hazardous materials that could, if released accidentally, affect large areas. Therefore, residents should participate in local town hall meetings or join Local Emergency Planning Committees (LEPCs) to stay informed about the types of hazardous materials present in their vicinity and the protocols set in place for accidental releases.
- **Hazard response plan:** Developing a hazard response plan is an essential strategy for managing these risks effectively. Quick response strategies involve delineating clear evacuation protocols and maintaining open lines of communication during emergencies. Make sure your family emergency plan includes strategies for handling hazardous waste.
- **Appropriate equipment:** Keep appropriate equipment in your emergency kit. Stocking an emergency kit includes gathering essential items like protective gear (masks and gloves), nonperishable food, water, flashlights, and batteries. For specialized protection during chemical exposure, simple yet effective items like duct tape and scissors can seal windows and doors to prevent airborne toxins from

entering the home. Ready.gov suggests incorporating specific items like scissors and plastic sheeting into emergency kits for better protection during shelter-in-place situations (*Chemicals and Hazardous Materials Incidents*, n.d.).

Regular drills and participation in community initiatives enhance household preparedness and raise awareness of broader risk management. Conducting frequent emergency drills helps ensure that family members know their roles and can execute plans without hesitation during real emergencies. These simulations should cover a wide range of scenarios, ensuring readiness for different types of incidents.

Education programs focusing on the prevention and mitigation of hazardous material accidents serve as powerful tools to reduce risks and impact. By learning safe handling practices and understanding warning signs of potential hazards, you can take proactive measures to prevent incidents. For instance, knowing how to identify signs of chemical leaks, such as unusual odors or visual warnings, enables quicker responses that can mitigate harm. Programs sponsored by organizations like PHMSA emphasize the importance of rigorous training and education, which significantly enhances overall safety (*Prevention/Mitigation Guidelines*, n.d.).

## Bringing It All Together

This chapter has explored the importance of customizing preparedness strategies to effectively handle diverse crisis situations. By analyzing geographic and environmental vulnerabilities, you can tailor your responses to specific threats, such as floods, earthquakes, or pandemics. This customization ensures that each household's strategy is aligned with the most pressing risks they might face, enhancing resilience and safety. Whether it's retrofitting homes for earthquake resistance or building communication plans during civil unrest, understanding local hazards is crucial for developing comprehensive emergency plans that address both immediate and long-term needs.

The chapter also emphasized practical measures like securing homes and creating family emergency plans, which are essential components of crisis management. These strategies, alongside exploring alternate power sources during outages and managing hazardous materials safely, empower you to maintain security and stability during unexpected events. This proactive approach protects families and also strengthens community bonds, ensuring everyone remains connected and supported during challenging times. Let's wrap up this book with the long-term maintenance of all the strategies you've learned throughout these chapters.

## 12

# LONG-TERM MAINTENANCE AND IMPROVEMENT

Ensuring the ongoing effectiveness and adaptiveness of your home's safety strategies is a crucial aspect of maintaining a secure household. It involves a continuous commitment to evaluating and enhancing the measures that protect your family and property from potential risks. As you, your family, and the environment around you change, so do the challenges you face, requiring constant vigilance and adaptability. This chapter sheds light on the importance of creating a proactive safety culture within your household that remains effective over time. By nurturing this approach, you empower yourself and your loved ones to respond swiftly and competently to any emerging threats, safeguarding physical structures and creating peace of mind.

This chapter uncovers various strategies that can help you keep your safety measures relevant and reliable for long-term usage. We will explore scheduled safety audits that systematically assess the integrity of household precautions, identifying potential hazards before they escalate. It emphasizes integrating these audits into routine schedules to ensure consistent evaluations. We'll also highlight the value of professional advice alongside DIY approaches, enriching the overall safety strategy with expert perspectives. In addition, it discusses maintaining an organized inventory of safety supplies, thereby ensuring readiness for emergencies and addressing personalized needs. Through technological advancements and continuous education, you can fortify your home to withstand evolving challenges, equipping families with the necessary tools to enhance long-term safety effectively.

### Scheduled Safety Audits and Updates

To create a safe home environment, it's valuable to do regular assessments that allow you to upkeep the security of your home. Routine audits reveal vulnerabilities that may otherwise go unnoticed. Over time, you and your family can become complacent, assuming that once safety measures are in place, they are effective indefinitely. However, household dynamics and environmental factors often change, necessitating frequent evaluations to maintain high levels of safety. Routine audits serve as essential tools for preventing this kind of complacency. By systematically examining each area of your home, you can identify potential hazards such as faulty wiring, loose handrails, or even just cluttered pathways that might pose risks. The following tips can help you successfully implement audits and systematic updates:

- **Add routine checks to your calendar:** To effectively implement these routine checks, integrating them into a structured calendar is advisable. Scheduling periodic safety audits makes them an organized part of your household priorities rather than random activities that are easily forgotten throughout daily routines. Consider setting aside specific days, perhaps once every season, to conduct comprehensive evaluations of your home's safety features. Doing this helps you instill a culture of continual vigilance, reinforcing the concept that safety is an ongoing responsibility rather than a one-time effort. This also helps distribute the workload evenly throughout the year, allowing for thorough inspections without overwhelming anyone.
- **Document your changes:** Documenting any changes or findings during these audits is another important aspect. Maintaining a detailed record serves multiple purposes: It provides a history of what has been done, helping track trends or recurring issues that need attention. This documentation proves invaluable when reassessing the effectiveness of past interventions, ensuring that no problem is addressed more than once unless necessary. Furthermore, keeping records aids in planning future updates and serves as an excellent reference during emergencies, enabling quick identification of shut-off points or critical contact numbers. As you document these observations, consider utilizing digital tools like apps or spreadsheets, which make retrieval easy and organization seamless.
- **Acquire professional assistance:** While you can perform many of these tasks independently, periodically enlisting professional assistance is wise. Experts bring fresh perspectives and advanced knowledge, catching potential issues that might escape the untrained eye. Professionals such as home inspectors or safety consultants have the expertise to uncover hidden safety concerns—like structural weaknesses or emerging electrical faults—that you might have overlooked. They come equipped with checklists and equipment designed for thorough evaluations, offering peace of mind and often suggesting improvements that can enhance overall safety significantly. To get the perfect balance and a well-rounded strategy, consider blending DIY approaches with professional insights. For instance, a professional inspection might reveal that older homes require updated wiring to accommodate modern appliances safely, or they may suggest additional insulation in areas prone to

dampness to prevent mold growth. These examples highlight how expert intervention can lead to meaningful upgrades and adjustments, elevating the standards of home safety far beyond what might be achievable through amateur assessments alone.

The benefits of involving professionals extend beyond immediate fixes. They also educate you on best practices and recent advancements in home safety technology. This knowledge transfer empowers you with the skills and understanding needed to maintain your home proactively. Engaging in conversations with experts provides valuable opportunities to learn about evolving threats and innovative solutions, ensuring that homeowners remain informed advocates for their families' safety.

Review and Replenish Supplies

The safety and well-being of your household relies on preparedness. To ensure readiness for any emergency, maintaining an adequate stock of safety and emergency items is paramount. Here are some tips you can consider when maintaining your supplies:

- **Organize inventory:** One essential aspect of effective preparedness is having an organized inventory. This approach makes checking supplies more efficient while preventing unexpected shortages that can arise during emergencies. Imagine a storm approaching and realizing your flashlight batteries are dead or missing. A scenario like this highlights the importance of meticulously organizing and tracking these vital resources.
- **Monitor stocks:** By keeping an organized inventory, you can easily monitor your stocks. This method allows you to quickly identify what needs replenishing before it becomes a concern. Some people choose to maintain a dedicated list, perhaps on a digital device or in a physical book. Others might opt for labeled storage containers to visually spot missing items at a glance. Regularly updating this inventory ensures that all safety materials are accounted for, supporting a household's resilience against unforeseen events.
- **Regularly check the emergency kit:** Another critical component of maintaining an emergency kit is regular checks on the expiration dates of stored items. Many safety and emergency supplies have a limited shelf life; for instance, food, medications, and even certain batteries can expire, diminishing their effectiveness when needed most. For example, canned foods may last years, but items like granola bars typically have shorter lifespans. Conducting frequent reviews of these dates helps guarantee that all provisions are safe and usable. In this setting, creating a schedule for these checks can be incredibly beneficial. Engaging family members in this task not only shares the responsibility but also educates everyone

involved. Scheduling these checks around significant calendar dates, such as the start of each season or daylight saving time changes, can serve as natural reminders. By doing so, you integrate these audits into your routine, making them less daunting and more habitual. Furthermore, replacing expired items promptly avoids the pitfall of finding yourself unprepared in a crisis.

- **Include personal variations:** Taking into account personal variations extends beyond medical or dietary considerations. Including favorite snacks or comfort items, especially for children, can offer emotional support during stressful times. It's also important to remember pets by setting aside enough food and water for them, along with comforting items like blankets or toys. Crafting personalized kits ensures the physical safety of all family members and also provides peace of mind, knowing that comprehensive preparations have been made.
- **Expand to the community:** In addition to personal efforts, community discussions about supplies are important to strengthen communal resilience. Engaging with neighbors or local groups to discuss emergency preparedness encourages a collective mentality toward safety. These conversations can lead to shared resources, mutual aid agreements, or even bulk purchasing of essential items, which can reduce costs. During power outages or severe weather, having a network of informed and prepared individuals can prove invaluable. This information can be shared during set community forums or meetings.

Ultimately, maintaining an adequate stock of safety and emergency items is not just about piling up supplies; it's about creating a thoughtful, strategic plan that encompasses organization, regular maintenance, personalization, and community engagement. When you dedicate attention to these areas, your household can achieve a level of readiness that secures your own safety while enhancing the overall resilience of your community. Preparedness is a continuous cycle that demands effort and diligence, yet its benefits during emergencies are immeasurable.

## Staying Informed on Emerging Threats

As with most things in life, threats evolve with time, so keeping up-to-date with these emerging threats can help you adjust your strategies to continuously keep your home safe. This sense of preparedness involves both individual efforts and collective intelligence through the following steps:

1. **Reliable sources:** The first step in fostering a culture of vigilance is relying on trusted sources for accurate information about potential risks. You should seek out reputable channels such as government websites, established news outlets, or specialized security advisories to stay informed about local threats or broader

security concerns. Regularly consulting these resources helps increase awareness of emerging issues, allowing you and your family to anticipate and mitigate risks before they become imminent threats.

2. **Community engagement:** Community alerts also significantly impact real-time updates that facilitate quick responses to potential dangers. These alerts can come from neighborhood watch groups, local law enforcement agencies, or emergency management offices. Joining or subscribing to community alert systems ensures you receive timely notifications about events like severe weather conditions, criminal activities, or health emergencies affecting your area. The immediacy of these alerts enables households to take proactive measures, such as securing properties, preparing emergency kits, or activating contingency plans when necessary. Active participation in community-based networks facilitates a shared responsibility for mutual safety and encourages collaboration among neighbors.

3. **Ongoing education:** Skill development through workshops further expands your ability to respond effectively to varying threats. Attending local workshops focused on emergency preparedness, self-defense, first aid, or fire safety equips individuals with practical skills and knowledge that can be life-saving during critical situations. Beyond personal development, these gatherings serve as networking opportunities where residents can share resources, exchange advice, and build supportive relationships with others who prioritize safety.

Encouraging vigilance and adaptability within your household involves engaging all family members in safety discussions and training. Holding regular family meetings to review security policies, discuss emergency procedures, and assign roles during crises inspires an inclusive environment where everyone feels empowered and responsible for their collective well-being.

Incorporating New Technologies and Innovations

Throughout this book, we've discussed the value of integrating technology into your practices, but how can you ensure you use devices and systems effectively and safely? The following steps can help you integrate cutting-edge innovations into home security strategies, allowing you to create a more secure environment that adapts to modern challenges:

- **Research new products:** The first step in enhancing home safety through technology is researching new safety products. The market is continually evolving, with companies developing innovative solutions designed to address emerging threats. Whether it's a smart lock that provides access control through biometric scanners or sensors that detect unusual water leaks and notify you via smartphone apps, staying informed about the latest technologies allows you to choose products that cater specifically to your needs. Dedicating time to research can help you discover solutions that are effective and cost-efficient in the long run.

- **Use smart technology in your home:** Smart technology has revolutionized home safety by integrating itself seamlessly with existing security systems. Devices such as cameras, motion detectors, and alarms can now be interconnected to create a comprehensive monitoring system that covers every corner of a property. For instance, smart cameras equipped with facial recognition can differentiate between familiar faces and potential intruders, alerting you instantly. Similarly, motion sensors can be programmed to activate lights when movement is detected outside specific hours, deterring potential threats. These devices often connect to cloud-based platforms, allowing you to monitor your property remotely, no matter where you are. This level of integration provides better protection and enhances convenience, enabling peace of mind with a glance at a smartphone.

- **Address security and safety concerns:** When discussing technology in home safety, it's essential to address privacy and security concerns. As smart devices become more prevalent, safeguarding personal data becomes crucial. You should ensure that the devices you choose adhere to strict security standards, utilizing end-to-end encryption and reliable authentication procedures. It's vital to select products from reputable manufacturers who prioritize user privacy. Being aware of the potential vulnerabilities in wireless connectivity, such as Wi-Fi or Bluetooth, helps take preemptive steps to safeguard against unauthorized access. Educating family members about safe Internet practices and regularly updating device firmware contribute to maintaining a secure technological environment at home.

- **Get expert help:** To further leverage technology in enhancing home safety, using professionals when necessary can provide insights that may not be apparent to the average user. Professionals can assess current systems, offering recommendations for upgrades or identifying weaknesses in existing setups. They bring expertise that can guide you in selecting technologies that meet your specific requirements, ensuring that investments translate into tangible safety improvements. Professional consultations can also introduce novel ideas and best practices, creating a proactive approach to home security.

Don't forget to continuously educate yourself on the technology you can use and the advantages behind the different types. Staying informed ensures that you can use the latest gadgets efficiently.

How to Create Checklists

If all else fails, resorting back to simple checklists can give you guidance and keep you accountable. Creating checklists can be a simple yet effective way to ensure your home is secure, especially in case of emergencies. It's important to think carefully about what you need to do before an emergency happens. Here are some steps to help you create checklists that will keep your house secure.

### *Step 1: Identify Potential Emergencies*

The first step in creating an effective checklist is identifying the types of emergencies you might face. Different locations have different risks. For example, if you live in an area prone to hurricanes, you will want a checklist specific to that situation. If you live in a region that experiences earthquakes, your checklist will look different. Take some time to think about what emergencies are most likely and make a list.

Once you have your list, categorize the emergencies into types. These could be natural disasters like floods or man-made situations such as break-ins. This categorization will help you create specific action points later on.

### *Step 2: Assess Your Home Security*

Next, take a thorough look at your home's current security measures:

1. Walk around your property both inside and outside.
2. Check the locks on doors and windows to make sure they operate correctly.
3. Look for any gaps where someone could enter without proper authorization.
4. Consider the lighting around your home. If an area is poorly lit, it might be easier for an intruder to go unnoticed.

Make a note of anything that could be improved. For example, if you find a window that does not close properly, include fixing it in your checklist. Additionally, think about investing in simple additions like motion-sensor lights or security cameras and add those to your list of improvements.

### *Step 3: Create a Checklist Format*

Now, it's time to decide how you want to format your checklist. Some people prefer a digital format, while others find paper checklists more convenient. Choose whatever works best for you. If you go digital, you could use a mobile app or just a simple document on your computer.

Consider organizing the checklist into sections based on the types of emergencies you identified earlier. For example, you could have one section for natural disasters and another for man-made threats. This way, when you need to reference it quickly, you can find the relevant information right away.

### *Step 4: Write Actionable Steps*

For each situation on your checklist, write down clear, actionable steps to take. If you listed a hurricane, this might include steps like gathering supplies. You should specify items needed, such as bottled water, nonperishable food, and a battery-operated radio. Be exhaustive but also concise.

For a break-in scenario, your checklist might include checking doors and windows, activating the alarm system if you have one, and informing neighbors to keep an eye on your home if you are away. Each step should be easy to understand so that anyone can follow it without confusion.

### Step 5: Review Emergency Contacts

An important aspect of home security involves having quick access to essential contacts. Include a section in your checklist for emergency contacts. This list should contain first responders, your local emergency management office, and family members. Keep this list updated, as numbers can change.

Consider adding a few local services that might be necessary during an emergency. For example, include contact information for your local plumber or handyman, as they can help with issues like burst pipes that could worsen during a storm.

### Step 6: Practice Your Plan

Creating a checklist is only one part of preparing your home for emergencies; you must also practice using it. Hold a family meeting to discuss the checklist and ensure everyone understands their responsibilities. This is critical for ensuring that everyone can react quickly in an emergency. You might want to run through a few scenarios where your family practices the steps together.

For instance, simulate what you would do in the event of a fire. Practice how to use fire extinguishers or how to safely exit the house. Knowing what to do can reduce panic during an actual emergency and help keep everyone safe.

### Step 7: Store Your Checklist Safely

Once your checklist is complete, think about where to keep it. You want it to be accessible but also safe from being damaged. A common strategy is to place a printed version on the refrigerator. This way, everyone can see it and refer to it regularly.

If you have opted for a digital version, ensure it is stored in a place that is easy to access, like a note-taking app that you always have on your phone. Whichever way you choose, the critical point is guaranteeing you and your family can find it when necessary.

### Step 8: Revise Periodically

Emergencies can evolve over time, so it is crucial to revisit your checklist periodically. Set a reminder every six months to review the checklist and make necessary updates. This could involve adding new emergency contacts or changing the supplies you need based on what you have used.

Additionally, if you've added new safety measures to your home or updated your security system, make sure those changes are reflected in your checklist to keep it accurate and useful.

### Step 9: Encourage Feedback From Family

Finally, encourage your family members to provide input on the checklist. They may have insights or suggestions on what to add. Perhaps they have ideas about additional precautions or have noticed security concerns you might not have considered.

Incorporating everyone's viewpoints can enrich your checklist and create a sense of shared responsibility. This process can help ensure everyone feels included and invested in emergency preparedness.

Taking the time to create effective checklists can significantly enhance your home's security. By following these steps, you can guarantee that you are well-prepared for various emergencies.

Final Thoughts

As we wrap up this chapter and this book, it's important to remember that the safety and security of your home should always come first. Regular audits and updates to your safety measures help reveal any new vulnerabilities, ensuring that your home remains a secure haven. By integrating routine checks into your schedules, involving professionals when necessary, and keeping organized records, you create a proactive approach to home safety. This prevents unexpected incidents and also allows for continuous improvement of your safety protocols.

Furthermore, staying informed about emerging threats and new technologies plays a crucial role in maintaining an updated safety strategy. Engaging with community resources, participating in workshops, and making use of modern technological tools all contribute to a reliable defense against potential risks. With open communication among household members and continuous learning, you ensure everyone is prepared and confident in handling emergencies. Your commitment to vigilance and adaptability solidifies your home's safety, providing peace of mind and protection for all family members.

# CONCLUSION

As we draw to a close, it's important to reflect on the journey we've undertaken together. Our exploration began with the understanding that creating a safe home involves more than just locking doors and windows. It's about a holistic approach that examines every potential vulnerability and implements solutions tailored to your specific needs.

Throughout this book, we've examined various strategies to bolster the security of your home, from fortifying physical barriers to establishing effective community networks. These strategies are not standalone measures as they are more effective when combined, forming a powerful safety net that protects your loved ones.

The essence of these lessons lies in their layered integration. Applying techniques such as strengthening exterior defenses, optimizing surveillance systems, and fostering open communication within your neighborhood creates a resilient environment capable of deterring all forms of threats. The value of this approach is its adaptability, allowing it to evolve as new challenges arise or circumstances change. This flexibility ensures that your efforts to enhance home safety remain relevant and effective over time.

Now equipped with essential tools and insights, it's time to translate them into action. The knowledge gained throughout these chapters is like a map leading toward a safer existence. You're encouraged to take immediate steps, no matter how small, to initiate this journey. Start by gathering your family for an important meeting—a safety audit where everyone participates in evaluating the strengths and weaknesses of your current setup. This collective engagement showcases areas that need improvement and creates a sense of shared responsibility among family members.

# CONCLUSION

Preparedness is not a destination but an ongoing voyage. As you venture this path, remember that it's crucial to maintain a proactive stance. Revisit your plans regularly, staying informed about new technologies or tactics that can further enhance your security. Consider preparedness as a living entity—nurture it consistently to ensure it grows alongside you. This mindset transforms preparedness from a daunting task into a natural part of daily life, seamlessly integrating itself into routines and family dynamics.

The significance of community involvement cannot be stressed enough. While individual efforts are foundational, collective action brings about substantial change. Imagine engaging with those around you, constructing a network of vigilant neighbors who look out for one another. In this context, developing initiatives, such as neighborhood watch programs, allows you to tap into a reservoir of resources and support that amplifies your own measures. Together, you create a culture where vigilance becomes second nature, reinforcing the safety of every household involved.

This sense of community extends beyond mere physical security. It's about building connections that promote open dialogue and mutual assistance. When crises arise, having established relationships ensures that help is readily available, reducing the burden on any single family and increasing resilience across the board. Collaborating with others means sharing knowledge, learning from diverse experiences, and collectively finding solutions that might otherwise remain elusive.

In addition to physical measures, the mental aspect of safety plays an equally important role. Resilience in the face of adversity is a skill that must be honed and nurtured. Just as you train muscles to grow stronger, mental resilience requires regular exercise. Engage in decision-making exercises with your family during calm times. Simulate scenarios where split-second choices may be necessary, allowing everyone to practice clear-headed thinking under pressure. This preparation ensures that when real emergencies occur, your responses are guided by confidence rather than paralyzed by fear.

Creating mental resilience also involves creating an environment where emotional expression is encouraged. Open discussions about fears, anxieties, and hopes related to safety contribute to psychological well-being. They provide opportunities for reassurance, enabling everyone to feel supported and understood. Knowing that there is space to express concerns validates each person's experience, strengthening bonds and reinforcing unity in the face of challenges.

As we conclude, let us reaffirm the importance of sustaining both physical and mental security. Your journey doesn't end here; in fact, it begins anew with each sunrise. With newfound knowledge and a commitment to safeguarding what matters most, you're better equipped to navigate the uncertainties of the future. Embrace this continuous cycle of learning, implementing, and adapting. Equip yourself and your loved ones with the tools to face whatever may come with courage and determination.

Through this dedication to preparedness, you can protect those you cherish and also inspire others to journey on similar paths. By leading through example, you instigate a ripple effect, encouraging the widespread practice of proactive measures that benefit everyone. This allows you to become an integral part of a movement that prioritizes safety and resilience, ensuring that families far and wide feel secure in their homes.

Let this conclusion serve as both a reflection and a launchpad. Reflect on the valuable insights gained and use them as stepping stones toward a safer tomorrow. Take pride in your efforts, knowing they contribute to a world where safety and peace of mind are within reach for everyone. Thank you for embarking on this journey. Together, we make a difference, one home at a time.

# REFERENCES

Abdelhadi, M. (2024, September 24). *Deadbolts vs. smart locks: Which is better for home security?* Low Rate Locksmith. https://www.lowratelocksmith.com/residential/security-alarms/deadbolts-vs-smart-locks-which-better-home-security/

*About neighborhood watch.* (n.d.). National Neighborhood Watch. https://www.nnw.org/about-neighborhood-watch

Ahrnsen, R. (2024). *Illuminate your home with these outdoor motion sensor lights, according to our tests.* Better Homes & Gardens. https://www.bhg.com/home-improvement/lighting/outdoor/best-outdoor-motion-sensor-lights/

Airhart, E. (2021, August 31). *The best emergency preparedness supplies.* The New York Times. https://www.nytimes.com/wirecutter/reviews/emergency-preparedness/

*Alternative currency: Substitute moneys and cryptocurrencies.* (2024, September 19). The Human Journey. https://humanjourney.us/economy/trust-faith-and-confidence-value-and-the-role-of-money/substitute-moneys-and-cryptocurrencies/

Andress, E., & Harrison, J. (n.d.). *Preparing an emergency food supply, short term food storage.* fcs.uga.edu. https://www.fcs.uga.edu/extension/preparing-an-emergency-food-supply-short-term-food-storage

Antosz, D. (2023, July 31). *What is a Neobank? How fintech is transforming banking.* Plaid. https://plaid.com/resources/fintech/what-is-a-neobank/

*Are you and your food prepared for a power outage?* (2021). USDA. https://www.usda.gov/media/blog/2016/09/26/are-you-and-your-food-prepared-power-outage

*Beneficial ownership information access and safeguards.* (2023). Federal Register. https://www.federalregister.gov/documents/2023/12/22/2023-27973/beneficial-ownership-information-access-and-safeguards

*Build a kit.* (n.d.). Ready.gov. https://www.ready.gov/kit

Bullock, J. A., Haddow, G. D., & Coppola, D. P. (2019). *Mitigation, prevention, and preparedness.* Introduction to Homeland Security. https://doi.org/10.1016/B978-0-12-415802-3.00010-5

Burke, T. (2024, April 26). *Fostering school safety through community involvement and collaboration.* Raptor Technologies. https://raptortech.com/resources/blog/fostering-school-safety-through-community-involvement-and-collaboration/

Canada, P. S. (2018, December 21). *Emergency preparedness guide for people with disabilities/special needs.* Get Prepared. https://www.getprepared.gc.ca/cnt/rsrcs/pblctns/pplwthdsblts/index-en.aspx

Celik, Y. (2022, May 9). *Why do you need an emergency management/evacuation plan?* First 5 Minutes. https://www.first5minutes.com.au/blog/why-do-you-need-an-emergency-management-evacuation-plan/

*Chemicals and hazardous materials incidents.* (n.d.). ready.gov. https://www.ready.gov/hazmat

Chris. (2024, June 21). Best fence for home security. *Pacific Fence & Wire Co.* https://www.pacificfence.com/blog/fencing-tips/best-fence-for-home-security/

*Complete guide to home safety assessment.* (n.d.). Spring Hills. https://www.springhills.com/resources/home-safety-assessment

Cooper, T. R. (2019, October). Creating a culture of preparedness. *Delaware Journal of Public Health.* https://doi.org/10.32481/djph.2019.10.003

*COVID-19 impacts leases while force majeure cases await decisions.* (2020). Pillsbury Law. https://www.pillsburylaw.com/en/news-and-insights/real-estate-force-majeure-covid-19-leases.html

*Crisis funding - how much do we really know?* (2021). Oxford Policy Management. https://www.opml.co.uk/insights/crisis-funding-how-much-do-we-really-know

Cutfarm_0jzjpr. (2024, April 3). *Comprehensive security guide for your home.* Pimlico Key Service. https://www.pimlicokey.com/archives/12279

# REFERENCES

Dahal, R., & Bista, S. (2023, February 20). *Strategies to reduce polypharmacy in the elderly*. PubMed; StatPearls Publishing. https://www.ncbi.nlm.nih.gov/books/NBK574550/

*Deadbolt installation: Everything you need to know*. (2024, May 23). Safety Locksmith. Safety Locksmith https://safetylocksmithbellevue.com/everything-you-need-to-know-about-deadbolt-installation/

Dev, R. E. (n.d.). *Small wind turbines & solar PV | Renewable off-grid energy systems*. Ryse Energy. https://www.ryse.energy/

Developing a disaster communication plan: 8 steps. (2023, January 18). *Tulane University*. https://publichealth.tulane.edu/blog/developing-disaster-communication-plan/

Dhaka, P., Bedi, J., Vijay, D., Singh Gill, J., & Barbuddhe, S. (2021). Emergency preparedness for public health threats, surveillance, modelling & forecasting. *Indian Journal of Medical Research*. https://doi.org/10.4103/ijmr.ijmr_653_21

*Disaster preparedness for children and youth with special health care needs*. (2020). Aap.org. https://www.aap.org/en/patient-care/disasters-and-children/professional-resources-for-disaster-preparedness/preparedness-for-children-and-youth-with-special-health-care-needs/?srsltid=AfmBOooY3_5RQFN2pWbQEIyWYA8JKTN93kTrT9jTjARzzCIo6h9PxVJj

*Does a clause in your commercial contract protect you during the COVID-19 crisis?* (2020, April 13). Cranfill Sumner LLP. https://www.cshlaw.com/resources/frustration-of-purpose-the-impossibility-of-commercial-contracts/

Editorial Contributors. (n.d.). *14 all-natural home remedies for fast cold & flu relief*. WebMD. https://www.webmd.com/cold-and-flu/14-tips-prevent-colds-flu-1

Edwards, R. (2022, February 22). *Complete home safety and security checklist*. SafeWise. https://www.safewise.com/checklists/home-security/

*8 easy ways to boost home security*. (n.d.). Home Gauge. https://www.homegauge.com/learning/boost-home-security/

*Electric power generation, transmission, and distribution*. (n.d.). osha.gov. https://www.osha.gov/etools/electric-power/illustrated-glossary

*Emergency backpack: 10 things you need*. (2023, August 30). Convoy of Hope. https://convoyofhope.org/articles/emergency-backpack/

*Emergency toolkit*. (2018, October 16). CARE Toolkit. https://www.careemergencytoolkit.org/management/7-project-management/7-budget-management/

*Empowering women: Urban self-defense and home security in challenging times*. (2024, October 2). The SurvivalTabs. https://thesurvivaltabs.com/blogs/news/empowering-women-urban-self-defense-and-home-security-in-challenging-times?srsltid=AfmBOoofomrW8Clw25QCUCOJVmce9zV1tsv0_93s8uXKQt15dcmAQYe5

*Evacuation procedures*. (2019). Csusm.edu. https://www.csusm.edu/em/procedures/evacuation.html

*15 Fam 670 building management programs and support*. (2024). State.gov. https://fam.state.gov/fam/15fam/15fam0670.html

Figuero, M. (2022, June 28). *11 Benefits of learning CPR*. AEDCPR. https://www.aedcpr.com/articles/11-benefits-of-learning-cpr/

Filip, R., Puscaselu, R. G., Anchidin-Norocel, L., Dimian, M., & Savage, W. K. (2022, August 7). Global challenges to public health care systems during the COVID-19 pandemic: A review of pandemic measures and problems. *Journal of Personalized Medicine*. https://doi.org/10.3390/jpm12081295

*Food and water safety during power outages and floods*. (2020, April 8). U.S. Food & Drug Administration. https://www.fda.gov/food/buy-store-serve-safe-food/food-and-water-safety-during-power-outages-and-floods

Forchuk, C., Serrato, J., Lizotte, D., Mann, R., Taylor, G., & Husni, S. (2022, April 29). *Developing a smart home technology innovation for people with physical and mental health problems: Considerations and recommendations*. JMIR MHealth and UHealth. https://doi.org/10.2196/25116

Forsetlund, L., O'Brien, M. A., Forsén, L., Mwai, L., Reinar, L. M., Okwen, M. P., Horsley, T., & Rose, C. J. (2021). *Continuing education meetings and workshops: Effects on professional practice and healthcare outcomes*. Cochrane Database of Systematic Reviews. https://doi.org/10.1002/14651858.cd003030.pub3

*Free home safety checks*. (2023). Arlingtonva.us. https://www.arlingtonva.us/Government/Departments/Fire/Safety/Free-Home-Safety-Checks

# REFERENCES

*Fueling dreams: Small business alternative lending strategies.* (2023). Loan Management Software by Fundingo. https://www.fundingo.com/category/api-integrations/page/3/

*Governor Ron DeSantis issues updates on state preparedness efforts for Hurricane Milton.* (2024). Flgov.com. https://www.flgov.com/2024/10/08/governor-ron-desantis-issues-updates-on-state-preparedness-efforts-for-hurricane-milton-3/

Hadley, C. (2020, September 18). *12 tactical experts share their everyday carry (EDC) gear & guns.* Tactical Hyve. https://tacticalhyve.com/everyday-carry-round-up/

Haughy, J. (2022, November 21). *Importance of CPR and first aid training.* Heart Start CPR. https://heartstartcpr.net/importance-of-cpr-and-first-aid-training/

*Home emergency preparedness: Creating a plan for natural disasters.* (2023, August 31). Stay Safe. https://staysafe.org/home-safety/home-emergency-preparedness/

*Home security audits: Assessing vulnerabilities and improving safety.* (2023, August 25). Stay Safe. https://staysafe.org/home-safety/home-security-audits-assessing-vulnerabilities-and-improving-safety/

*How effective are home security systems?* (n.d.). adt.com. https://www.adt.com/resources/home-security-system-effectiveness

*How secure are deadbolts?* (2023). Artie's Locke & Key. https://artieslockandkey.com/2023/08/how-secure-are-deadbolts/

*How to build a strong family support system.* (2024, August 17). Fei Yue. https://fycs.org/how-to-build-a-strong-family-support-system/

*How to maintain your emergency kit.* (n.d.). REI. https://www.rei.com/learn/expert-advice/how-to-maintain-your-emergency-kit.html

*Impact of family engagement.* (2021). Youth.gov. https://youth.gov/youth-topics/impact-family-engagement

Juvare Staff. (2024, September 24). *Mutual aid in emergency management: A growing trend.* Juvare. https://www.juvare.com/mutual-aid-in-emergency-management-a-growing-trend/

Kennedy, R. (2023, March 16). *Top 3 must-have tactical gear for survival.* Chase Tactical. https://www.chasetactical.com/guides/top-3-tactical-gear-for-survival

Kubala, J. (2018, February 4). *18 remedies to get rid of headaches naturally.* Healthline. https://www.healthline.com/nutrition/headache-remedies

LeWine, H. E. (2024, April 3). *Understanding the stress response.* Harvard Health Publishing. https://www.health.harvard.edu/staying-healthy/understanding-the-stress-response

Ma, C., Qirui, C., & Lv, Y. (2023, December 14). *"One community at a time": Promoting community resilience in the face of natural hazards and public health challenges.* BMC Public Health; BioMed Central. https://doi.org/10.1186/s12889-023-17458-x

*Maintaining vigilance to combat terrorism.* (n.d.). National Institute of Justice. https://nij.ojp.gov/topics/articles/maintaining-vigilance-combat-terrorism

*Make a first aid kit.* (2023). American Red Cross. https://www.redcross.org/get-help/how-to-prepare-for-emergencies/anatomy-of-a-first-aid-kit.html

Makki, M., Hassali, M. A., Awaisu, A., & Hashmi, F. (2019, June 13). *The prevalence of unused medications in homes.* Pharmacy. https://doi.org/10.3390/pharmacy7020061

Mărcuță, C., & MoldStud Research Team. (2024, June 13). *Mobile app development for emergency response.* Moldstud. https://moldstud.com/articles/p-mobile-app-development-for-emergency-response

Mayo Clinic Staff. (2018). *First-aid kits: Stock supplies that can save lives.* Mayo Clinic. https://www.mayoclinic.org/first-aid/first-aid-kits/basics/art-20056673

*Mindfulness STOP Skill.* (n.d.). Cognitive Behavioral Therapy Los Angeles. https://cogbtherapy.com/mindfulness-meditation-blog/mindfulness-stop-skill

*Multihazard emergency planning for schools toolkit.* (n.d.). Training.fema.gov. https://training.fema.gov/programs/emischool/el361toolkit/siteindex.htm

*Neighborhood watch programs: Collaborating for enhanced security.* (2023, August 28). Stay Safe. https://staysafe.org/home-safety/neighborhood-watch-programs-collaborating/

Nick, G. A., Savoia, E., Elqura, L., Crowther, M. S., Cohen, B., Leary, M., Wright, T., Auerbach, J., & Koh, H. K.

# REFERENCES

(2009). *Emergency preparedness for vulnerable populations: People with special health-care needs.* Public Health Reports. https://www.ncbi.nlm.nih.gov/pmc/articles/PMC2646456/

Nortje, A. (2020, July 1). *10+ best grounding techniques and exercises to strengthen your mindfulness practice today.* Positive Psychology. https://positivepsychology.com/grounding-techniques/

*Off-grid or stand-alone renewable energy systems.* (2020). U.S. Department of Energy. https://www.energy.gov/energysaver/grid-or-stand-alone-renewable-energy-systems

Ozbay, F., Johnson, D. C., Dimoulas, E., Morgan, C. A., Charney, D., & Southwick, S. (2007, May). *Social support and resilience to stress: From neurobiology to clinical practice.* Psychiatry (Edgmont); Matrix Medical Communications. https://www.ncbi.nlm.nih.gov/pmc/articles/PMC2921311/

*Performance under pressure (how to manage stress).* (2017, March 13). First10EM. https://first10em.com/performance-under-pressure/

*Personal protective equipment.* (2023). Occupational Safety and Health Administration. https://www.osha.gov/personal-protective-equipment

Pogue, T. (2023, November 15). *Safety first: Outdoor lighting tips for a secure home.* Resort Lighting Inc. https://resortlightinginc.com/home-safetyhome-securityinformative-blog/safety-first-outdoor-lighting-tips-for-a-secure-home/

*Post evacuations.* (2024, August 13). United States Department of State. https://www.state.gov/global-community-liaison-office/crisis-management/post-evacuations/

*Preparing an emergency food supply, long term food storage.* (n.d.). fcs.uga.edu. https://www.fcs.uga.edu/extension/preparing-an-emergency-food-supply-long-term-food-storage

*Prevention/mitigation guidelines.* (n.d.). phmsa.dot.gov. https://www.phmsa.dot.gov/grants/hazmat/preventionmitigation-guidelines

*Protective actions research.* (2024). Fema. https://community.fema.gov/ProtectiveActions/s/article/Power-Outage-Sign-Up-for-Alerts-and-Warnings

*Public safety and crisis communication in an emergency or disaster.* (n.d.). Rural Health Info. https://www.ruralhealthinfo.org/toolkits/emergency-preparedness/3/public-safety

*Relaxation techniques for stress relief.* (2018, October 23). Help Guide. https://www.helpguide.org/mental-health/stress/relaxation-techniques-for-stress-relief

*The role of lighting in enhancing outdoor safety and security.* (n.d.). American National Sprinkler & Lighting. https://americannationalco.com/the-role-of-lighting-in-enhancing-outdoor-safety-and-security/

Rosen, M. A. (2019). *Teamwork in healthcare: Key discoveries enabling safer, high-quality care.* American Psychologist; NCBI. https://doi.org/10.1037/amp0000298

Samardzic, M., Doekhie, K. D., & Wijngaarden, J. D. H. (2020). *Interventions to improve team effectiveness within health care: A systematic review of the past decade.* Human Resources for Health. https://doi.org/10.1186/s12960-019-0411-3

Schauf, C. (2018). How to purify water in the wild. *Uncharted Supply Company.* https://unchartedsupplyco.com/blogs/news/purify-water-in-wild?srsltid=AfmBOorBiEJWU-U1x8P9N2KG5cYk1n-5lvPTitWW34YZJe8bJEt1zqZn

*School emergency alert notification and crisis alert management system.* (n.d.). Catapult Emergency Management. https://www.catapultemergencymanagement.com/

Schuman-Olivier, Z., Trombka, M., Lovas, D. A., Brewer, J. A., Vago, D. R., Gawande, R., Dunne, J. P., Lazar, S. W., Loucks, E. B., & Fulwiler, C. (2020). *Mindfulness and behavior change.* Harvard Review of Psychiatry. https://doi.org/10.1097/HRP.0000000000000277

*Section 3 - asset management - Part I: Investment principles, policies and products.* (2024). Fdic.gov. https://www.fdic.gov/bank-examinations/section-3-asset-management-part-i-investment-principles-policies-and-products

*Shelf-stable food safety.* (2023). Usda.gov. http://www.fsis.usda.gov/food-safety/safe-food-handling-and-preparation/food-safety-basics/shelf-stable-food

Shukla, M., Amberson, T., Heagele, T., McNeill, C., Adams, L., Ndayishimiye, K., & Castner, J. (2024, May 1). Tailoring household disaster preparedness interventions to reduce health disparities: Nursing implications from machine learning importance features from the 2018–2020 FEMA National household survey. *International Journal of Environmental Research and Public Health, 21*(5), 521. https://doi.org/10.3390/ijerph21050521

Tarlengco, J. (2024, June 27). *What is PPE in safety?* Safety Culture. https://safetyculture.com/topics/ppe-safety/

# REFERENCES

Taylor, M. (2022, April 28). *What does fight, flight, freeze, fawn mean?* WebMD. https://www.webmd.com/mental-health/what-does-fight-flight-freeze-fawn-mean

Thomas, J. (2020, December 5). *Building up a well-stocked pantry & long-term food storage supply*. Homesteading Family. https://homesteadingfamily.com/building-up-your-long-term-food-storage-supply/

*Tips for maintaining and upgrading survival equipment*. (2024, August). Tactical Survival Solutions. https://tacticalsurvivalsolutions.com/maintaining-equipment/

Top fences to add security to your home. (2024). *Ergeon*. https://www.ergeon.com/blog/post/security-fence-for-home

*Top 10 communication methods in a disaster setting*. (n.d.). Adjusters International. https://www.adjustersinternational.com/resources/news-and-events/top-10-communication-methods-in-a-disaster-setting/

Torani, S., Majd, P. M., Maroufi, S. S., Dowlati, M., & Sheikhi, R. A. (2019, April 24). The importance of education on disasters and emergencies: A review article. *Journal of Education and Health Promotion*. https://doi.org/10.4103/jehp.jehp_262_18

Tuohy, J. P., & Lopez, S. (2024). Why get home security cameras? *U.S. News & World Report*. https://www.usnews.com/360-reviews/services/security-cameras/why-get-security-cameras

*Uncertainty principles in crisis decision-making*. (2024). Fiveable.me. https://fiveable.me/quantum-leadership/unit-11/uncertainty-principles-crisis-decision-making/study-guide/XrOpm5xaXjVklgYD

Wallace, L. (2023, October 29). *The importance of community vigilance in cyber security*. The Missing Link. https://www.themissinglink.com.au/news/vigilance-cyber-security

*Water purification for emergency survival*. (n.d.). Emergency Kits. https://www.emergencykits.com/emergency-water/water-purification

*What to do if someone breaks into your home*. (n.d.). adt.com. https://www.adt.com/resources/what-to-do-if-someone-breaks-into-your-house

*What is a home safety assessment?* (2023, September 5). Measurabilities. https://measurabilities.com/what-is-a-home-safety-assessment-2/

Wilson, C., Hargreaves, T., & Hauxwell-Baldwin, R. (2017, April). *Benefits and risks of smart home technologies*. Energy Policy. https://doi.org/10.1016/j.enpol.2016.12.047

Wolf-Fordham, S., Curtin, C., Maslin, M., Bandini, L., & Hamad, C. D. (2015). Emergency preparedness of families of children with developmental disabilities: What public health and safety emergency planners need to know. *Journal of Emergency Management, 13*(1), 7-18. https://doi.org/10.5055/jem.2015.0213